Effectiveness of Disinfecting Wastewater Treatment Plant Discharges: Case of Chemical Disinfection Using Performic Acid

Effectiveness of Disinfecting Wastewater Treatment Plant Discharges: Case of Chemical Disinfection Using Performic Acid

Edited by

Vincent Rocher and Sam Azimi

Published by   **IWA Publishing**
Republic – Export Building, 1st Floor
2 Clove Crescent
London E14 2BE, UK
Telephone: +44 (0)20 7654 5500
Fax: +44 (0)20 7654 5555
Email: publications@iwap.co.uk
Web: www.iwapublishing.com

First published 2021
© 2021 IWA Publishing

Disclaimer

British Library Cataloguing in Publication Data
A CIP catalogue record for this book is available from the British Library

ISBN: 9781789062090 (paperback)
ISBN: 9781789062106 (eBook)

Contents

Preface ix

Contributors xi

General introduction xxi

**Section 1: Effectiveness of Chemical Disinfection at the
Laboratory Scale** 1

Chapter 1
*Laboratory scale study of wastewater disinfection by
means of PFA and the factors affecting its effectiveness* . . . 3

1.1   Introduction 3
1.2   Trial Methods Used at the Laboratory Scale 4
1.3   Comparison of PFA Disinfection Effectiveness with
      other Chemicals 7
1.4   Influence of PFA Dose on the Effectiveness of Fecal
      Bacteria Removal 9
1.5   Influence of WWTP Discharge Quality on Fecal
      Bacteria Removal 12
1.6   Impact of PFA Application in Partially Treated Wastewater on
      PFA Dose Requirements 15
1.7   PFA Effectiveness in Removing other Microorganisms 19
Key Points 21

Chapter 2
Fecal bacteria regrowth and viability after disinfection with PFA 22

2.1   Introduction 22
2.2   Experimental Description 23
2.3   Results 26
Key Points 29

Chapter 3
Impact of PFA on organic matter and post-injection consequences 30

3.1   Introduction 30
3.2   Impact of PFA on the Soluble Organic Matrix 31
3.3   Study of PFA Instability after Injection 35
Key Points 38

Chapter 4
The fate of micropollutants and byproduct formation during the disinfection of WWTP discharge by PFA 39

4.1   Introduction 40
4.2   Targeted Screening of Disinfection Byproducts 43
4.3   Non-Targeted Investigation of Micropollutants During PFA
      Disinfection 49
Key Points 55

Section 2:   Effectiveness of the Chemical Disinfection Process at Full Scale (Seine Valenton WWTP) 57

Chapter 1
Technical description of the industrial trials conducted at Seine Valenton WWTP 59

1.1   Introduction 59
1.2   Presentation of the Seine Valenton WWTP 60
1.3   Presentation of PFA Disinfection by Application of the Kemira
      KemConnect DEX Technology 61
1.4   Design of the Industrial-Scale Trials 61
1.5   Analytical Parameters Monitored 62

Chapter 2
In situ monitoring of fecal bacteria at the Seine Valenton WWTP using ALERT rapid microbiology instrumentation (Fluidion®) **64**

2.1 Introduction 64
2.2 Experimental Description 68
2.3 Results of a Side-by-Side Laboratory Comparison 70
2.4 Comparison of *in situ* Results on WWTP Effluent Disinfection 73
Key Points 73

Chapter 3
Effectiveness of PFA disinfection implemented at full scale (Seine Valenton WWTP) **75**

3.1 Introduction 75
3.2 Effectiveness of PFA Disinfection Applied to SEV WWTP Discharge 76
3.3 Mathematical Correlation 81
3.4 Interactions Between PFA and the Physicochemical Quality of SEV WWTP Discharge 88
Key Points 89

Section 3: Assessment of the Eco-Toxicological Effects of Disinfection Processes on the Seine River **91**

Chapter 1
Description of the biological models used to monitor water quality **93**

1.1 Description of the Biological Panels Used to Assess General Toxicity 93
1.2 Description of the Biological Panels Used to Assess Endocrine Disruption 96

Chapter 2
Biological models applied to the case of chemical disinfection using PFA **100**

2.1 Toxicity Assessment at the Laboratory Scale 100
2.2 Toxicity Assessment at the Industrial Scale 106
Key Points 111

Section 4:   Feedback from other Municipalities on the use of Chemical Disinfection with PFA **113**

Chapter 1
Case of Biarritz (France) **115**
Thierry Pigot and Thomas Paulin

1.1   Introduction 115
1.2   Materials and Methods 117
1.3   Effectiveness of Fecal Bacteria Disinfection 119
1.4   Environmental Impact of the Disinfected Effluent 120
Key Points 123

Chapter 2
Case of Venice (Italy) **124**
Patrizia Ragazzo and Nicoletta Chiucchini

2.1   Introduction 124
2.2   Materials and Methods 126
2.3   Results 128
Key Points 134

General conclusion **135**

References **137**

List of Figures **159**

List of Tables **201**

Preface

Writing a book is not an easy task, as anyone who has ever written one knows. Besides the content, there are many questions to ask before successfully telling the story … A book to say what? For whom? In what form? And once you have all the answers to these questions, there is one last question that remains elusive: How do you tell the story? Raymond Queneau illustrates this difficulty very well by telling us in 99 different ways the simple story of a man traveling in a bus (Exercices de style, Raymond Queneau. 1947).

Regarding the subject of access to water, its use, and more particularly, on the subject of bathing in urban areas, the choice and the way of telling the story are even more complicated. There are not 99 but thousands of ways to approach the subject, thousands of ways to tell this story of primary interest to each of us. Each of these ways will be fair, each of these ways will answer many questions and fuel reflections … But each of these ways will also leave out a number of questions. To choose means to exclude.

As a public authority in charge of wastewater treatment, our role has been to tell the story of disinfection from a technical perspective. A technology has been chosen and tested, from laboratory scale to full-scale trials… From an idea to its industrial feasibility. We tried to answer all the questions that came to a stakeholder's mind, with an operator and an end-user eye. No bias on the results. No position taken on the technology. The objective of our story has been to give, in a factual way,

© IWA Publishing 2021. Effectiveness of Disinfecting Wastewater Treatment Plant Discharges: Case of Chemical Disinfection Using Performic Acid
Editors: Vincent Rocher and Sam Azimi
doi: 10.2166/9781789062106_ix

the results of the tests, starting from the development of the protocols, passing through the evaluation of technology's effectiveness and the verification of its harmlessness on the environment. Thus, the reader can, at will, read all or a part of the book according to his own questions, expectations, and objectives ... To tell his own story!

In order to be able to address all of these subjects, to answer all of these questions, we have put all the required skills around the table, ending up all together with 23 co-authors. This method of doing things is limitlessly rich since collective competence is much greater than the sum of individual skills and that's why all co-authors are named at the beginning of the book regardless to their specific contribution to each chapter. Only the fourth part includes the name of the contributors at the beginning of each chapter since these works have been added to the book to complete the overview.

We hope that you will have as much pleasure in reading this book as we have had in writing it...

Happy reading!

Vincent Rocher and Sam Azimi

Contributors

The editors

Vincent Rocher heads up Innovation at the Greater Paris Sanitation Authority (SIAAP). After completing his environmental studies with a thesis on the fate of micropollutants in the sewer system, he entered SIAAP to work as a member of the technical and scientific support team. As Director of Innovation, Vincent now works on defining and implementing SIAAP research and innovation policy, in overseeing: asset management aspects, operational control and optimization of the entire sanitation system, and preparation of technical specifications and future regulatory requirements. Leader of the scientific research program, he provides support to the Executive Board and all SIAAP operational departments on technical and scientific matters.

Sam Azimi jointly heads up Innovation at the Greater Paris Sanitation Authority (SIAAP). Following his studies with a thesis on pollutant transfer through the atmosphere in urbanized areas, Sam worked as Operations Manager at a wastewater treatment plant within the SIAAP jurisdiction. He commissioned and operated the plant for eight years before joining the Innovation Division. His main objective is to preserve the link between Research and Development activities in the field of wastewater treatment and the needs of plant operators to optimize their processes. His primary tasks are twofold: coordinate SIAAP's scientific research program, and provide expertise on the water treatment processes and environmental impact of SIAAP's activities.

Sam Azimi and Vincent Rocher

The editors would like to thank the co-authors, the editorial supervisor and other contributors of this book, details of whom are given below.

The co-authors

GREATER PARIS SANITATION AUTHORITY (SIAAP)

Romain Mailler is an environmental engineering scientist and head of an R&D Service at SIAAP's Innovation Division. He was awarded a PhD in Environmental Science and Technology in 2015. His work has focused on gaining an understanding of all types of sewage mechanisms. In particular, Romain is presently working on the coagulation/flocculation of wastewater, oxidation and adsorption techniques, the fate of micropollutants in sewage facilities, sludge rheology and thickening/dewatering, and membrane/concrete/metal aging in wastewater. He is also a co-leader of the research mission dedicated to material aging in wastewater facilities as part of the MOCOPEE program.

Sabrina Guérin-Rechdoui is an environmental engineering scientist and head of an R&D Service at SIAAP's Innovation Division. Holder of a Master's degree in environmental engineering, she oversees development and deployment of innovative measures (physicochemical and biological) on environmental matrices. Sabrina is also in charge of the quality observatory for the Seine River in the Paris Metropolitan Area. She joined SIAAP in 2009 and began her career in process engineering. Extremely involved in research activities, she co-leads the MOCOPEE program's research mission on metrology and signal processing, in addition to being a member of the MeSeine Innovation program coordination committee.

Perrine Mèche is a research technician in charge of the technical platform dedicated to reagents used in wastewater at SIAAP's Innovation Division. Over the last

10 years she has conducted various studies on innovative treatment processes and reagent efficiency and characterization, spanning the laboratory, pilot and industrial scales. Perrine is currently working on mechanisms to better understand and optimize the various processes requiring the addition of reagents: coagulation/flocculation, biological treatments, and tertiary treatments by oxidation or adsorption, including wastewater disinfection.

Sébastien Pichon is a research technician in charge of the technical platform dedicated to process engineering studies at SIAAP's Innovation Division. He earned a D.U.T. professional degree in Biological Engineering (Lille University of Sciences and Technologies) in 1999 and prior to his current experience, Sébastien has conducted various studies in the field of wastewater treatment spanning the laboratory, pilot and industrial scales. He had also worked on developing a microbiology laboratory specialized in wastewater treatment and moreover conducted environmental and health studies related to the microbiological quality of water in cooling circuits.

Angélique Goffin is an environmental scientist at SIAAP's Innovation Division. She earned her PhD in Environmental Science and Technology in 2017. Before her SIAAP experience, she worked at the University of Paris Est-Créteil, in the Laboratoire Eau, Environnement et Systèmes Urbains (LEESU). Her focus lies in dissolved organic matter behavior in various aqueous environments. Angélique is also actively involved in developing multidimensional fluorescence spectroscopy with applications in surface water monitoring, water treatment and wastewater treatment. She received an award in 2018 for accomplishments in fluorescence spectra decomposition during the first international contest devoted to 3D fluorescence spectra in a parallel factor analysis.

Jean Bernier is an R&D engineer and head of an R&D Service at SIAAP's Innovation Division. He holds a PhD in Water Engineering, obtained in 2013 from Laval University, Québec. His work has primarily focused on modeling wastewater treatment processes/plants and river quality, as well as using these models in practical applications for SIAAP. In particular, Jean has worked on biofiltration, physicochemical primary settling, anaerobic digestion and membrane bioreactors. His work also entails automated data processing and sensor measurement validation. In addition, he is co-leader of the MOCOPEE program's research mission on modelling and process control in the area of wastewater.

Jérôme Roy is the Director of three SIAAP wastewater treatment plants. A graduate of ESTP Paris combined with a Master's degree in Environmental Engineering and Management from the ENSMP School, he began his career in the construction industry and quickly transitioned into the environmental protection sector.

He then spent nearly 15 years supervising the building of wastewater treatment plants for SIAAP. Since 2016, Jérôme has held the post of Deputy Director and then Director for the three SIAAP wastewater treatment plants located in the eastern part of the Greater Paris area, upstream along the Seine River. This unique geographic configuration makes these WWTP plants critical to hosting open water events at the Paris 2024 Olympic Games.

PARIS-EST CRETEIL UNIVERSITY

Gilles Varrault is a Professor of chemistry at the Laboratoire Eau, Environnement et Systèmes Urbains (LEESU) at the University of Paris Est Créteil. He obtained his PhD in the field of environmental sciences from the University of Paris VII in 2001. He was appointed to the post of Assistant Professor at the University in 2002 and received tenure in 2013. Until 2014, his research activities had mainly pertained to metal and organic pollutants in aquatic systems, with special attention focused on their bioavailability. Since 2014, Gilles has also worked on characterizing organic matter in receiving environments as well as in wastewater treatment plants through the use of spectrometric techniques. Furthermore, he helps develop sensors for the high-frequency, *in situ* monitoring of organic matter.

Julien Le Roux has been an Assistant Professor at the Laboratoire Eau, Environnement et Systèmes Urbains (LEESU) at the University of Paris Est-Créteil since 2015. He earned an engineering degree and PhD in Water Chemistry and Microbiology from the University of Poitiers (France) in 2008 and 2011. Before joining LEESU, Julien worked as a post-doctoral fellow at KAUST University (Saudi Arabia). He possesses expertise in water quality and treatment (especially oxidation processes), analytical chemistry (gas and liquid chromatography coupled to mass spectrometry, high-resolution mass spectrometry) for the analysis of organic micropollutants, identification of unknown compounds and elucidation of degradation mechanisms.

Nina Huynh holds an engineering degree from the Ecole Nationale Supérieure de Chimie in Rennes (France) and a Master's degree in water quality and treatment from the University of Poitiers (France). During her current PhD curriculum at the University of Paris Est-Créteil, Nina is involved in characterizing organic micropollutants in urban waters by means of high-resolution mass spectrometry.

Maolida Nihemaiti is currently a research scientist at the Helmholtz Centre for Environmental Research (UFZ) in Leipzig, Germany. She received a PhD from Curtin University in Perth, Australia in 2018. Before joining UFZ, she worked for one year at the University of Paris Est-Créteil with the Campus France Prestige Postdoctoral Research Program. Maolida's research interests include the fate of trace organic contaminants in various natural and engineered aquatic environments and an analysis of their transformation products through the use of high-resolution mass spectrometry.

PAU AND PAYS DE L'ADOUR UNIVERSITY

Thierry Pigot is a tenured professor at the University of Pau and Pays de l'Adour and conducts research at the Institute of Analytical Sciences and Physio-chemistry for the Environment and Materials (IPREM UMR CNRS 5254). He heads the Federation of Research on Environments and Aquatic Resources, which brings together several research teams working in the field of water and aquatic environments. Thierry's particular research topics concern the study of chemical and photochemical oxidation processes and mechanisms applied to water and air treatment. He is also involved in several Bachelor's and Master's degree courses in the following fields: environmental chemistry, water treatment, organic chemistry.

Thomas Paulin is a research engineer at the Institute of Analytical Sciences and Physio-chemistry for the Environment and Materials (IPREM UMR CNRS 5254), currently responsible for IPREM's Chemical and Microbiological Water Analysis Laboratory in Anglet. Thomas was the operations manager during the study conducted on the performic acid disinfection process at the Biarritz wastewater treatment plant, where he coordinated all monitoring campaigns and environmental studies.

SORBONNE UNIVERSITY

Jean-Marie Mouchel is a Professor at the Milieux Environnementaux, Transferts et Interactions dans les hydrosystèmes et les Sols (METIS) unit at the Sorbonne University (Paris). As a trained hydrologist and biogeochemist, he led for 10 years the PIREN-Seine program, a vast project focusing on the functioning of the Seine River basin, viewed as a socio-eco-system. Jean-Marie presently heads the UMR 7619 Metis joint research laboratory (Sorbonne University, CNRS, EPHE) in the field of environmental sciences, whose teams are specialized in hydrology, near-surface geophysics, nutrient and carbon fate and cycling, and contamination studies. His primary research topics are dedicated to water quality, particularly in densely populated areas.

WATCHFROG LABORATORY

David Du Pasquier is a senior scientist at the Watchfrog Laboratory. He holds a PhD in Molecular Endocrinology and is specialized in endocrine disruption. He joined the company in 2006 to work on the development and use of *in vivo* tests using frog and fish embryos for the detection of endocrine disruption in water samples, cosmetics or chemicals. David is a member of the OECD validation management group for ecotoxicity testing and sits on the OECD Chemical Committee. In these expert groups, he participates in the development and international validation of tests used for the regulatory assessment of chemicals with respect to endocrine disruption.

TAME-WATER

Laurent Paulic is a microbiologist with 15 years of experience at Tame-Water. He developed the company's technical product offering and oversaw its implementation with customers. From his initial training in the food industry, he worked in the pharmaceutical industry before focusing on plasma physics for 10 years in the field of controlling atmospheric contamination. A business developer who went on to become Laboratory Supervisor and President, Laurent's company was subsequently acquired by the Alcen Group. Since that time, he has been expanding the product side of the business, including the development and deployment of automated biosensors.

FLUIDION S.A.

Dan Angelescu is the CEO and R&D Director of Fluidion, founded in 2012. The company's activities are focused on novel technology approaches to the *in situ* chemical and microbiological analysis of water, wastewater and the environment. After completing his BS from the California Institute of Technology and his PhD from Princeton University (both in Physics), he worked as a senior researcher at Schlumberger Technologies, where he helped develop the first downhole microfluidic analytics platform. He then continued his career in academia, in being awarded the French Research Accreditation. As a Professor at the University of Paris-Est, Dan led a research group focused on the interplay between nanotechnology and microfluidics, with specific emphasis on novel sensors. In parallel, he taught classes in science and engineering disciplines, including a co-leadership position in an international Master's program. He has authored the book 'Highly Integrated Microfluidics Design' along with numerous peer-reviewed articles and patents in the areas of condensed matter physics, nanotechnology and microfluidics.

Vaizanne Huynh is an Operations Engineer at Fluidion. Following her Master's degree in Organic Chemistry, she undertook a Master's in Innovation at the Sorbonne University in Paris, where she joined the company FLUIDION for her Master's thesis in 2014. She initiated development of the pH sensor prototype and then, as an R&D Engineer, participated in the company's microbiology sensor project. Vaizanne oversaw the metrological part of the FLUIDION microbiological sensor, known as the ALERT System. Since its release on the market, she has stepped into the role of Operations Engineer handling implementation, client follow-up and other projects. She has also co-authored several peer-reviewed articles in the fields of organic chemistry and microbiology.

Andreas Hausot is the team's Director of engineering at Fluidion. He holds a B.A. Sc. degree in Civil Engineering from the University of British Columbia (Vancouver, Canada) and is author of multiple patents and patent applications on technologies spanning oilfields, microfluidics, subsea monitoring and sampling. He has previously worked as Project Manager for downhole fluid analysis at several international locations with Schlumberger, leading engineering teams in developing sensors for extreme environments, with an emphasis on optical fluid analysis technology. Andreas is a co-founder of Fluidion and moved to France to become the team's Director of Engineering, focusing on the development of innovative new technologies for autonomous water quality measurements.

VERITAS S.p.A.

Patrizia Ragazzo heads the R&D–EU projects unit for the Integrated Water Service (IWS) of VERITAS S.p.A, the multi-utility managing Venice and Venice region's coastal areas, which encompass some one million residents. Before working with VERITAS, she was a researcher in the Chemical Engineering Department of Padua University and ran the R&D-Process Control Department of A.S.I. S.p.A, an IWS utility operating in northeastern Italy. Patrizia's field of research spans both drinking water and wastewater processes, having published some 80 papers in national/international journals and proceedings. Disinfection is one of her primary specializations.

Nicoletta Chiucchini is a chemical engineer at the R&D–EU projects unit of VERITAS S.p.A Integrated Water Service. Prior to this experience, she held the post of Assistant Manager for the R&D-Process Control Department of A.S.I. S. p.A, an IWS utility operating in northeastern Italy. Her work has mainly focused on wastewater treatment and technologies (management, optimization, upgrade, etc.) and related research activities to enhance processes. Effluent disinfection is among her areas of specialization.

The editorial supervising

Géraldine Izambart, head of the innovation management unit at SIAAP's Innovation Division and **Pauline Souchal**, from her team carried out the whole editorial supervising of the present book.

The contributors

In addition to the co-authors there are a number of contributors to this book who have provided significant assistance, in particular for the full scale trials detailed in part 2.

Marc Hamon (SIAAP Seine Amont – HQSE) and Fabrice AUBE, André Lévy and Bruno Kembellec (SIAAP DSAR – Combined sewer system operators) who participate in implementing the sampling equipment.

Sylvain Fouillaud and Sébastien Lahaie (SIAAP Seine Amont – laboratory) operation and maintenance of sampling equipment.

Jean-Baptiste Tonnerieux (SIAAP Seine Amont – Project manager), Xavier Letallec and Hafida Hadjami (SIAAP Seine Valenton – Plant process team) for the coordination of the project and the fieldwork.

Sébastien Riello and Julien Gervais (SIAAP Seine Valenton – Plant operators) for the adapted operation of the plant to fit with the need of the project and all the data treatment.

Jean-Louis Thénard, Mathieu Paillard (DSAR – PC Saphyr Effluent regulation managers) and their team for the adapted combined sewer operation and the effluent regulation during the trials.

Nicolas Gillon, Victoria Deng and Marc Fontaine (DSAR – Operation optimization unit) for their help to adapt effluent flows.

Moreover, gratitude is expressed to Eden Galot (an SIAAP intern), as well as to the teams at the SIAAP Laboratory for their meaningful contribution to the experimental campaign of this project (Part 1 – Chapter 1). The authors would also like to thank Patricia-Aubeuf-Prieur from Kemira for her assistance with the laboratory-scale PFA fabrication protocol and full-scale trials (Parts 1 and 2). The authors would like to thank the PRAMMICS platform and OSU-EFLUVE (UMS 3563) for providing access to the UPLC-IMS-QTOF, and specifically to Emmanuelle Mebold for her assistance during the analyses (Part 1 – Chapter 4). These analyses was supported by a public grant overseen by the French National Research Agency (ANR) as part of the JCJC program (WaterOmics project, ANR-17-CE34-0009-01). M. Nihemaiti received funding from the People Programme (Marie Curie Actions) of the European Union's Seventh Framework Programme (FP7/2007-2013) under REA grant agreement no. PCOFUND-GA-2013-609102, through the PRESTIGE program coordinated by Campus France. The development of the ALERT System (Part 2 – Chapter 2) was partially funded through the '*Prototypes Technologiques*' program sponsored by the Paris Region. The financial and technical support from the City of Biarritz and the entire Basque Metropolitan Zone has been greatly appreciated (Part 1 – Chapter 2 and Part 4 – Chapter 1) and the technical assistance offered by wastewater treatment plant staff (SUEZ) is also gratefully acknowledged.

General introduction

Over the past 50 years, the layout of the Parisian sanitation system has radically changed, shaped through three major milestones. Beginning in the early 1970s and over a 10-year period, a major industrial revolution led to a strong treatment capacity expansion at the Seine-Aval wastewater treatment plant, which would become one of the largest plants worldwide. This first step was followed by an another crucial 10-year phase, from 1980 to 1990, which focused not only on building new plants to bring on-line a significant increase in treatment capacity but also, and based on an intensive applied research activity, on preparing for the treatment plant changes required in response to environmental expectations. Lastly, the changes occurring during the past 25-plus years, from 1990 to present day, have been aimed at improving water treatment quality from removing only carbon to the complete removal of carbon, nitrogen and phosphorus (Rocher & Azimi, 2017). Such industrial changes have significantly contributed to decreasing pollutant discharges into the Seine River while restoring its physiochemical quality. Consequently, the oxygen level in the river is now quite good, and the nitrogen and phosphorus concentrations are much lower than in past years. These conditions favor the resurgence of numerous fish species, with a recent record of 32 (whereas only three were counted in 1970) (Rocher & Azimi, 2016).

The Parisian sanitation system now faces new challenges in managing the wastewater produced by nine million inhabitants in a densely urbanized area. More efficient use and management of wastewater are critical to addressing the growing demand for water and the threats to water security. In this context, the

© IWA Publishing 2021. Effectiveness of Disinfecting Wastewater Treatment Plant Discharges: Case of Chemical Disinfection Using Performic Acid
Editors: Vincent Rocher and Sam Azimi
doi: 10.2166/9781789062106_xxi

evolution of social expectations in favor of wastewater reuse and the City of Paris' pledge to open the Seine River to Olympic and Paralympic swimmers, as well as to all Parisians, in 2024 have brought the issue of bacterial pollution to the forefront (Directive 2006/7/CE, 2006). This issue seeks to identify and quantify the main input pathways of bacteria into the river (treatment plants, storm overflows, etc.) in order to define the best action plan for achieving the recreational use objectives. Various solutions have been devised on the scale of the Paris Metropolitan Area, and scenarios have been tested using the PROSE mathematical model to establish and prioritize the actions to be carried out. PROSE is a modeling tool, calibrated and validated on the Seine River, intended to simulate nutrients and bacterial contamination within the Parisian watershed; moreover, since 2007, it has been used by the Greater Paris Sanitation Authority (SIAAP) to design the Greater Paris sanitation master plan.

For all scenarios examined, disinfection of the effluent conveyed by the wastewater treatment plants located upstream of the City of Paris appears to be a necessary action. Indeed, bacterial contamination of the Seine River has been shown to be significantly higher downstream of the wastewater treatment plant outfalls (Figure 1).

During dry weather conditions, wastewater treatment plants significantly contribute to bacterial concentrations in the river. The median concentration discharged into the river has ranged from 5×10^3 to 50×10^3 MPN/100 mL of *Escherichia coli* (*E. coli*) (Rocher & Azimi, 2016), leading to a median concentration level in the river of 1200 MPN/100 mL at the Port à l'Anglais site, located downstream of the outfall. Thus, to be compliant with the bathing quality objectives, the disinfection of the outfall becomes crucial to limit the bacterial concentrations in the river. The situation is quite different during rain events. Relying mostly on a combined sewer system, the Greater Paris Sanitation Authority operates more than 400 km of pipes to handle wastewater for treatment into six plants. While during dry weather periods the facilities are able to treat the entire flow, such is not always the case during wet weather periods. During rain events, due to the high level of impermeable surfaces, additional rainwater significantly increases the flow rate into sewer pipes and exceeds the drainage and treatment capacities of the whole sewer system, thus inducing combined sewer overflows (CSOs) into the river. This untreated CSO discharges water, containing up to 5.5×10^6 MPN/100 mL of *E. coli*, into the river (Rocher & Azimi, 2016). In this case, CSOs become the main introduction pathway of *E. coli* into the Seine River (Figure 1, yellow curve, occurring during the rain event after 28 hours of simulation) and the disinfection of the plant outfall, which was of prime importance during dry periods, appears useless for the river to be compliant with bathing quality objectives

While the main objectives for the years ahead concern the ability to manage rainwater in compliance with the water framework directive (WFD, 2006), tackling the bacterial contamination due to wastewater during dry periods becomes of critical importance in significantly reducing the Seine River contamination level. Though Rocher and Azimi (2016) have shown that conventional wastewater treatment facilities offer a satisfactory bacteria removal rate (ranging from 2.76 to 3.03 log for *E. coli* and 2.73 to 3.37 log for intestinal enterococcus), it would appear that additional treatment processes are still required in order to significantly reduce the bacterial contamination of WWTP outfalls.

To help stakeholders in their choice for any additional treatment, the SIAAP has integrated into its scientific program, through its Mocopée research program, an action to study the disinfection processes and their effectiveness and harmlessness to the environment.

Among the technologies currently available for disinfecting wastewater outfalls, chemical disinfection appears to be a well-adapted solution due to both its operational flexibility and efficacy at removing bacterial contamination. While several chemicals are known to disinfect water, performic acid (PFA) is currently being used by some municipalities (Biarritz, France and Venice, Italy). This is the basis for SIAAP's testing of PFA to disinfect its Seine Valenton wastewater treatment plant outfall. A two-year study was performed both in the laboratory and at full-scale. This provided answers regarding both the efficacy and the absence of environmental impact of this disinfection process. This book sets out and discusses the expected outcomes of using PFA under three headings:

– Assessment of the efficacy of this type of chemical disinfection on WWTP outfall;
– Technical feasibility of full-scale implementation;
– Verification of the minimal impact on the River Seine.

A fourth section, which reports feedback from the municipalities where PFA disinfection has already been implemented, serves to complete this overview.

Section 1

Effectiveness of Chemical Disinfection at the Laboratory Scale

Laboratory scale study of wastewater disinfection by means of PFA and the factors affecting its effectiveness

© SIAAP

1.1 INTRODUCTION

The Paris Metropolitan Area intends to authorize recreational bathing in the Seine River during the summer by 2024. The precondition to achieving this objective consists of limiting as much as possible the input of fecal bacteria into the river, as per current bathing regulation targets for *Escherichia coli* (*E. coli*) and intestinal enterococci (*Directive 2006/7/EC, dated 15 February 2006 concerning the management of bathing water quality, repealing Directive 76/160/\EEC,*

doi: 10.2166/9781789062106_0003

2006). One necessary (although insufficient) action is to disinfect the wastewater treatment plant (WWTP) discharge upstream of Paris.

In the case of the Paris Metropolitan Area, performic acid (PFA), peracetic acid (PAA), UV irradiation and ozonation seem to be among the disinfection techniques existing for wastewater application, effective options when considering how they respond to fecal bacteria. In addition, the application of chlorination processes has not been considered since the disinfection unit would be implemented to treat wastewater treatment plant (WWTP) effluents discharged into the Seine River.

PFA has recently been tested at both the laboratory scale (Chhetri *et al.*, 2014; Gehr *et al.*, 2009; Karpova *et al.*, 2013; Luukkonen *et al.*, 2015) and industrial scale (Chhetri *et al.*, 2015; Ragazzo *et al.*, 2013, 2017) in order to disinfect treated or partially treated wastewater. This technique would offer several operational advantages, for example, effectiveness at a relatively low dosage, lack of harmful byproducts, an onsite production just before injection and post-injection instability (Karpova *et al.*, 2013; Luukkonen *et al.*, 2015; Ragazzo *et al.*, 2013). In addition, such a technique could be easily activated during summer only (bathing period) and then switched off the rest of the year. Therefore, a multidisciplinary project was conducted covering the Paris Metropolitan Area in order to study the possibility of implementing a PFA disinfection unit at a plant within the SIAAP jurisdiction.

The first objective of this project was to study, at the laboratory scale, the wastewater disinfection by PFA and the factors affecting its effectiveness, and then to validate this choice before carrying out industrial-scale tests. Various experiments were thus performed to: (1) validate the choice of PFA in comparison with PAA; (2) quantify precisely the relationships between $C \times t$ (concentration $\times$ time of exposure) and fecal bacteria reduction in the targeted WWTP discharge; (3) assess the effect of water quality, especially particles, on PFA effectiveness; (4) evaluate the feasibility and implications of disinfecting partially treated wastewater by PFA; and (5) obtain information regarding the effect of PFA on other types of pathogens. This first chapter will present all the experiments and results associated with these five points.

1.2 TRIAL METHODS USED AT THE LABORATORY SCALE
1.2.1 PFA preparation method

PFA is produced at the laboratory scale in accordance with the two-step Kemira preparation protocol described in Luukkonen (2015). The first step consists of catalyzing formic acid by adding 10% by mass of sulfuric acid. This step is performed in an iced batch since the reaction is exothermic. The second step entails producing PFA from the catalyzed formic acid by addition into an iced bath of hydrogen peroxide with a mass ratio of 1/1. A maturation time during mixing is needed given that the reaction is slow and the mass concentrations of hydrogen peroxide and PFA change until reaching equilibrium. Specific experiments have been conducted to determine an optimal maturation time of around 1 hour. In theory, at the end of preparation, the solution contains 20% hydrogen peroxide and 13.5% PFA. The solution is always controlled by dosing the exact PFA concentration. This method is summarized in Figure 2. In theory, the final solution containing 13.5% of PFA also contains 30.9% of residual formic acid, hence the injection of 1 ppm of PFA leads to injecting 0.79 ppm of C (0.19 from PFA plus 0.60 from formic acid). The PFA solution is then directly stored at $-18°C$ in darkness for a maximum period of 3 weeks.

1.2.1.1 PFA titration method

The reference dosing technique consists of a redox colorimetric titration of hydrogen peroxide in the solution mixed with diluted sulfuric acid by adding potassium permanganate. The equivalency is determined by a persistent pink coloration of the solution. Next, potassium iodide is added, which leads to transforming PFA into a carboxylic acid along with the formation of 1 mole of iodine for each mole of carboxylic acid. Lastly, the iodine is dosed using thiosulfate in the presence of starch, leading to discoloration of the solution upon reaching equivalency. This method is summarized in Figure 2.

1.2.1.2 Decay kinetics at different temperatures of PFA produced at the laboratory scale

PFA preparation and titration at the laboratory scale have been tested in order to evaluate: (1) repeatability of the titration technique; (2) repeatability of the preparation protocol; and (3) decay kinetics of PFA at various storage temperatures (-18, 4 and 20°C) in darkness.

Regarding repeatability of the titration technique, five titrations of a given PFA solution were performed on the same day. This evaluation was also conducted on different days and with different PFA solutions. The average concentrations obtained for PFA and hydrogen peroxide were respectively: 11.73 $\pm$ 0.80 and 19.79 $\pm$ 0.65%. The variability of this method was therefore very low, with a coefficient of variation equal to 7 and 3%. As regards repeatability of the laboratory scale preparation protocol, five PFA solutions were prepared and dosed using the colorimetric method over five different days. For samples 3, 4 and 5, titration was performed three times. The PFA concentrations obtained were: 12.12% (sample 1), 11.91% (sample 2), 12.40 $\pm$ 0.98% (sample 3), 13.03 $\pm$ 0.91% (sample 4), and 11.24 $\pm$ 0.28% (sample 5). The overall average obtained was thus 12.18 $\pm$ 0.94%, for a limited variability of just 8% (coefficient of variation), which is comparable to the titration method variability. The laboratory scale PFA preparation protocol is hence reliable and repeatable.

As for decay kinetics, PFA concentration was monitored in one of the preparations stored in darkness at three different temperatures (-18, 4 and 22°C).

Figure 3 displays the evolution in PFA concentration vs. time over a period of 150 days.

In the three cases studied, the PFA decay follows a pseudo-first-order kinetics ($y = A\,e^{-k\,x}$). The influence of temperature is clear, with an increase in constant k as temperature increases: 0.004 at $-18°C$, 0.041 at $4°C$, and 0.16 at $22°C$. PFA exhibits special behavior when stored at $-18°C$, with a concentration rise during the first 3 days (28%), followed by a pseudo-first-order decay kinetics. The initial PFA concentration is then once again reached after 20–30 days. It can be assumed that PFA formation continues when the solution is kept at a temperature below $0°C$. The half-life of PFA is thus 210, 17 and 4.5 days, respectively, at -18, 4 and $22°C$. In more practical terms, if PFA use is considered to remain the same until its concentration reaches 90% of the initial concentration, then a PFA preparation can be stored for a maximum time of 63, 2.5 and 0.5 days, respectively, at -18, 4 and $22°C$. Out of safety concerns, a PFA preparation should be titrated for each use, as was the case for this work.

1.2.1.3 Description of an experimental PFA disinfection method at the laboratory scale

Laboratory scale disinfection trials were performed in 2-L batches under strong mixing conditions (200 rpm) in order to guarantee batch uniformity. A Jar-test apparatus, which is normally used for coagulation-flocculation batch tests, was employed herein: 1.8 L of the wastewater sample is first introduced into the batch; then, mixing is started and a given volume of the solution containing the disinfectant, corresponding to the targeted dose, is injected. After the contact time, 20 mg/L of thiosulfate is added to stop the disinfection reaction, mixing is stopped and the samples removed for analysis. For all laboratory scale disinfection trials, a 10-min contact time was applied. This method is summarized in Figure 2.

Most trials were conducted at the Seine Valenton (SEV, Valenton, France) WWTP, with distinct discharge samples being drawn several hours before the trials. SEV is run by SIAAP and treats 600,000 m^3/day of wastewater from the eastern part of the Paris Metropolitan Area. In a normal configuration (dry weather), the treatment processes are pre-treatment (screening, grit and oil removal), followed by a primary settling unit and biological treatment by an extended aeration activated sludge process to achieve a complete treatment of carbon, nitrogen (nitrification and denitrification) and phosphorus. A physicochemical dephosphatation unit treats a portion of the treated water in order to raise the level of phosphorus removal. In a degraded mode of WWTP operations (wet weather or internal bypass due to insufficient treatment capacity), the excess wastewater is conveyed directly to the physicochemical dephosphatation unit for particle and phosphorus removal. WWTP discharge to the Seine River depends on whether the water has been totally treated or is a mix of totally and partially treated. A complete description of SEV WWTP operations is given in Section 2, Chapter 1.

Various parameters were measured during most of these laboratory scale experiments, i.e. total suspended solids (TSS), dissolved organic carbon (DOC), chemical (COD) and biochemical (BOD) oxygen demands, total Kjeldahl nitrogen (TKN), total phosphorus (TP), and *E. coli* and intestinal enterococci (in MPN/100 mL). These analyses were performed according to the following methods: NF EN 872 for TSS, NF EN 1484 for DOC, ISO 15,705 for COD, NF EN 1899 for BOD, NF EN 25,663 for TKN, NF EN ISO 6878 for TP, NF EN ISO 9308-3 for *E. coli*, and NF EN ISO 7899-1 for intestinal enterococci.

Moreover, preliminary tests were conducted to consider the best analytical method for the quantification of the ions NH_4^+, NO_3^-, NO_2^- and PO_4^{3-} since the conventional method using cadmium column was not adapted. Indeed, the potential presence of residual PFA concentrations may cause damages to the column. These analyses were performed according to the following methods: NF EN ISO 11732 for NH_4^+, NF EN ISO 10304-1 for NO_3^- and NO_2^- and NF EN ISO 15681-2 for PO_4^{3-}. A significant impact from excess thiosulfate on the quantification of COD, BOD and DOC had been observed; these parameters were then analyzed in sub-samples free of thiosulfate.

1.3 COMPARISON OF PFA DISINFECTION EFFECTIVENESS WITH OTHER CHEMICALS

1.3.1 Description of the experiments performed

For starters, PFA was compared to peracetic acid (CAS 79-21-0) and hypochlorite (CAS 7681-52-9), both of which are commonly used disinfectants. An initial literature survey did lead to identifying PFA and peracetic acid (PAA) as promising disinfection alternatives for wastewater (Chhetri *et al.*, 2014, 2015, 2018; Luukkonen and Pehkonen, 2017; Luukkonen *et al.*, 2015; Ragazzo *et al.*, 2013). These initial experiments were aimed at both comparing their disinfection effectiveness on SEV WWTP discharge and estimating the overall dose required of these compounds to achieve a given fecal bacteria elimination rate. Hypochlorite was adopted as the disinfection reference even though its application in the environment had not been considered.

Peracetic acid (VWR 1.07222.1000, mass concentration of 38–40%) and hypochlorite (Nectra 156031, mass concentration of 10%) were commercial solutions available from VWR, whereas PFA was directly produced at the laboratory scale following the preparation previously described (Figure 1). The applied PFA solution was produced and dosed on the same day as the disinfection experiments in order to precisely determine the volume of PFA to inject.

The three products were tested simultaneously on six distinct SEV WWTP discharge samples, except for PFA at doses of 5, 8 and 10 ppm, which were only tested on the first three samples. The treated wastewater was sampled a few hours before each disinfection trial (Nov 14, Nov 28, Dec 5, Dec 19 and Dec 20, 2017 and Jan 15, 2018). The previously described laboratory scale experimental disinfection method was applied. For each trial, four disinfectant doses were tested simultaneously (2, 5, 8 and 10 ppm) with a contact time of 10 min. The quality of the wastewater samples was evaluated before a disinfection step aimed at respecting the quality standards described in Part 2.1.3. The overall quality of the six SEV WWTP discharge samples was normal, with average concentrations of: 6.5 $\pm$ 2.8 mg/L for TSS, 5.7 $\pm$ 1.0 mgC/L for DOC, 21 $\pm$ 4 mgO_2/L for COD, 2.1 $\pm$ 1.4 mgO_2/L for BOD_5, 1.4 $\pm$ 0.5 mgN/L for TKN, and 1.1 $\pm$ 0.8 mgP/L for TP. Similarly, the *E. coli* and intestinal enterococci concentrations in the samples were close to what is generally encountered in this type of water with median $\pm$ min-max concentrations of: 13,288 $\pm$ 1690–123,603 and 5887 $\pm$ 1929–23,655 MPN/100 mL, respectively (Rocher & Azimi, 2016).

1.3.1.1 *Effectiveness of PFA, PAA and hypochlorite on SEV WWTP discharge*

Figure 4 Presents the results of the six disinfection trials performed to compare PFA, PAA and hypochlorite on SEV WWTP discharge with a contact time of 10 mins. Depicted here are the median concentrations of *E. coli* or intestinal enterococci, with the observed minimum and maximum concentrations shown as error bars. These values were calculated from six trials, except for PFA at 5, 8 and 10 ppm (three trials).

These trial results indicate a lower fecal bacteria disinfection effectiveness of PAA compared to PFA at a comparable limited applied dose. At an applied dose of 2 ppm, the residual concentrations of *E. coli* and intestinal enterococci in the disinfected wastewater are in fact equal to 38 $\pm$ <38–78 MPN/100 mL (median $\pm$ min-max) for PFA, and 2625 $\pm$ 1855–23,695 and 3596 $\pm$ 1511–19,701 MPN/100 mL, respectively, for PAA. This difference in effectiveness at 2 ppm is significant (Mann–Whitney test) for both *E. coli* and intestinal enterococci with p-values of 0.004. A higher dose is thus required to achieve a given fecal bacteria elimination rate with PAA compared to PFA. It can also be concluded that PFA disinfection effectiveness is comparable to hypochlorite at the same applied dose even though fecal bacteria concentrations after 10 min of disinfection at 2 ppm are slightly higher for hypochlorite than PFA. As an example, the residual *E. coli* concentrations in the disinfected wastewater are: 38 $\pm$ <38–78 MPN/100 mL (median $\pm$ min-max) for PFA, and 108 $\pm$ <38–2253 MPN/100 mL for hypochlorite. The difference in effectiveness at 2 ppm for these products is insignificant (Mann–Whitney test) for both *E. coli* and intestinal enterococci, with *p*-values of 0.177.

Karpova (2013) observed similar trends in concluding that 0.5 ppm of PFA had the same effectiveness in reducing fecal coliform concentration in biologically-treated wastewater as 1 ppm PAA and 8 ppm Cl_2. The lower PFA dose requirement compared to other chemicals is commonly acknowledged in the scientific literature (Luukkonen & Pehkonen, 2017). More recently, Ragazzo (2020) reported that for a typical disinfection contact time (20–30 min) at doses set to guarantee the current effluent limit for *E. coli* of 5000 CFU/100 mL, PAA and hypochlorite required Ct values three and four times higher on average, respectively, than PFA.

These preliminary results highlight the consistency of using PFA to disinfect the SEV WWTP discharge since this chemical features high fecal bacteria disinfection effectiveness at relatively low doses, which would in fact limit the quantity of chemical injected into the Seine River, in addition to the advantage of being produced *in situ* due to its instability.

1.4 INFLUENCE OF PFA DOSE ON THE EFFECTIVENESS OF FECAL BACTERIA REMOVAL

1.4.1 Description of the experiments performed on SEV WWTP discharge

Given the high disinfection effectiveness of PFA with a limited applied dose and time, multiple disinfection trials were performed on SEV WWTP discharge at various doses and a contact time of 10 min in order to study the dose-effectiveness relationships for both *E. coli* and intestinal enterococci. The aim herein was to both determine the required PFA C × t to achieve a given fecal bacteria elimination rate and characterize the variability due to water quality variations.

In all, 24 distinct SEV WWTP discharge samples were collected for PFA disinfection trials, including the six previously presented, between Nov 14, 2017 and Feb 13, 2019. The previously described laboratory scale experimental disinfection method was applied. The SEV WWTP discharge was sampled several hours before each disinfection trial and various PFA doses were applied depending on the sample, resulting in nine values for the 0.2–0.3 ppm interval, 13 values for 0.35–0.50 ppm, nine values for 0.7–0.8 ppm, 16 values for 0.9–1.0 ppm, eight values at 1.2 ppm, 15 values for 1.8–2.0 ppm, and lastly three higher values at 4.5, 7.2 and 8.9 ppm of PFA. The wastewater sample quality was evaluated before a disinfection step aimed at respecting the quality standards described above. The overall quality and variability of the 24 SEV WWTP discharge samples were relatively normal, with average concentrations of: 8.7 ± 5.5 mg/L for TSS, 5.8 ± 0.8 mgC/L for DOC, 23 ± 8 mgO$_2$/L for COD, 2.0 ± 1.2 mgO$_2$/L for BOD$_5$, 1.1 ± 0.7 mgN/L for TKN, and 1.2 ± 0.7 mgP/L for TP. Similarly, the *E. coli* and intestinal enterococci concentrations in the samples were close to what is normally encountered in this type of water, with median $\pm$ min-max concentrations of: $14{,}290 \pm 1690$–678 750 and 4630 ± 707–62,969 MPN/100 mL, respectively (Rocher & Azimi 2016). The variability of water quality originates from differing SEV WWTP operating conditions during the sampling period (internal bypass, dry and wet weather, etc.).

1.4.1.1 Chart of PFA effectiveness for the SEV WWTP discharge

Figure 5 displays the results of the 24 disinfection trials performed with PFA on SEV WWTP discharge with a contact time of 10 min; it depicts the median concentration of *E. coli* or intestinal enterococci, with the minimum and maximum concentrations observed as error bars (upper part of the figure), and the average logarithmic removal of both fecal bacteria, with the standard deviation shown as error bars (lower part of the figure). The quality limits (900 MPN/100 mL for *E. coli* and 330 MPN/100 mL for intestinal enterococci) for respecting European recreational bathing regulations (*Directive 2006/7/EC dated 15 February 2006 concerning the management of bathing water quality, repealing Directive 76/160/|EEC 2006*) are listed as indications along with the bacteria limit of quantification (LOQ).

The strong influence of the PFA dose at a contact time of 10 min is clearly observable. The increase in the applied PFA dose leads to an increase in fecal bacteria reduction until doses of between 1.8 and 4.5 ppm (C × t of 18–45 ppm. min), at which point the median concentrations of both bacteria reach the limits of quantification for this type of treated wastewater. Moreover, the PFA effectiveness is higher for *E. coli* than for intestinal enterococci, leading to a lower PFA dose requirement for these bacteria. This greater sensitivity of *E. coli* to PFA is particularly observable at low PFA doses, with more pronounced decreases of concentration from one dose to another for *E. coli* and higher logarithmic removal values. For example, the median *E. coli* concentration is lowered from 14,290 MPN/100 mL in the raw water to 179 MPN/100 mL in the disinfected water at 0.35–0.5 ppm PFA, while it is only reduced from 4630 to 1305 MPN/100 mL for intestinal enterococci at this same dose. The logarithmic removals are always higher for *E. coli* than for intestinal enterococci when considering doses below 1.8–2.0 ppm PFA, at which point the limits of quantification start to be reached. Such is particularly the case for a PFA dose of 0.35–0.5 ppm, with logarithmic removals of 0.66–2.85 and 0–1.71 respectively for *E. coli* and intestinal enterococci. Above this dose, the difference in logarithmic removals is due to higher initial concentrations for *E. coli* and similar limits of quantification. In the scientific literature, a comparable effectiveness of

PFA in disinfecting treated wastewater has been mentioned. Karpova (2013) obtained a removal rate of about 1 log for *E. coli* at 4–5 ppm.min of PFA applied to biologically-treated wastewater effluents at the laboratory scale, and 0.5–2.0 log for intestinal enterococci. Similarly, Luukkonen (2015) obtained, at the laboratory scale, a removal of 3.3 log for *E. coli* with an initial concentration of 29,200 CFU/100 mL and 15 ppm.min of PFA, compared to a removal of 2.5 log for intestinal enterococci with an initial concentration of 1840 CFU/100 mL. Next, industrial-scale trials for the disinfection of WWTP discharges with PFA were performed in Venice (Ragazzo *et al.*, 2013), concluding that 23 ppm.min or less of PFA led to *E. coli* and intestinal enterococci removals of respectively 2–4.2 and 0.7–3.2 log, while 60 ppm.min guaranteed removals always remained higher than 3 log. In these trials, the higher sensitivity of *E. coli* to PFA was also observed.

In addition, a strong variability in residual concentration and logarithmic removals was observed for a given applied PFA dose. For example, the residual *E. coli* and intestinal enterococci concentrations varied respectively between <15 and 1541 MPN/100 mL and between 38 and 20,488 MPN/100 mL for a PFA dose of 0.35–0.5 ppm, and between <15 and 419 MPN/100 mL and between <15 and 2720 MPN/100 mL for a dose of 0.9–1.0 ppm. Consequently, the logarithmic removal rate at a given PFA dose is also strongly variable, that is, approximately 1 log removal of variation from one sample to another. This variability is most likely indicative of an impact of the SEV WWTP discharge quality variations on PFA disinfection effectiveness. Among the 24 samples, 12 were in fact collected during nominal SEV WWTP operations (i.e., nominal flow capacity not being exceeded and no internal bypass), as opposed to the other 12, for which the SEV WWTP was experiencing partially degraded operations (i.e., flow capacity exceeded and/or internal bypass). This discrepancy led to variations of: TSS between 2 and 25 mg/L, COD between 6 and 42 mgO_2/L, DOC between 3.9 and 7.0 mgC/L, TKN between <0.3 and 2.8 mgN/L, NH_4^+ between <0.3 and 1.3 mgN/L, NO_3^- between 13.1 and 18.9 mgN/L, NO_2^- between <0.02 and 0.32 mgN/L, and PO_4^{3-} between <0.1 and 7.4 mgP/L.

Hence, in the case of SEV WWTP discharge disinfection, a C × t of 9–12 ppm. min seems to be a good target since within this range, the concentrations of *E. coli* and intestinal enterococci most of the time lie below the European recreational bathing regulation targets of 900 and 330 MPN/100 mL, respectively. As regards regulating the applied PFA dose in recognition of both the contact time (flow rate) and water quality, a target of 10 ppm.min could be considered for 'good quality' water and 20 ppm.min in the event of quality degradation. However, the quality parameters leading to a variability in results should be better understood in order to correctly modulate the PFA dose injection as a function of water quality.

1.5 INFLUENCE OF WWTP DISCHARGE QUALITY ON FECAL BACTERIA REMOVAL

1.5.1 Normalization of the PFA effectiveness to the SEV WWTP discharge quality

As observed in the previous section, a strong variability in PFA disinfection effectiveness has been observed, along with a variability in the SEV WWTP discharge quality. To identify suspected causes of degraded quality parameters, statistical correlation tests (Spearman, $\alpha = 0.05$) were performed between water quality parameters and the residual bacterial concentration or logarithmic removal at a PFA C × t of either 7–9 ppm.min ($n = 15$) or 9–12 ppm.min ($n = 18$). For both cases, no significant correlations (p-value >0.05) were found between the logarithmic removals of both types of bacteria and TSS, COD, DOC and initial bacterial concentrations. For a PFA C × t of 7–9 ppm.min, the residual *E. coli* concentration was statistically correlated with TSS ($r = 0.773 - p$-value $= 0.001$), COD ($r = 0.542 - p$-value $= 0.039$) and the initial *E. coli* concentration ($r = 0.703 - p$-value $= 0.005$), while the sole statistically correlated parameter with the residual intestinal enterococci concentration was COD ($r = 0.572 - p$-value $= 0.028$). For the higher PFA C × t of 9–12 ppm.min, the sole statistically significant correlation was found between the residual concentration of *E. coli* and TSS ($r = 0.637 - p$-value $= 0.006$). This finding would indicate that the PFA disinfection effectiveness in reducing *E. coli* and intestinal enterococci contents depends on the initial concentrations of bacteria, TSS and COD of the water.

In considering this conclusion, the results of the 24 disinfection trials were plotted in different ways in Figure 6, which represents the residual concentration of both types of bacteria after 10 min of disinfection vs. the normalized PFA C × t. The normalization step was performed by dividing the applied C × t by both initial bacterial concentrations, TSS or COD since they are correlated with the residual bacterial concentration after disinfection.

For one thing, Figure 6 highlights that the normalization steps are consistent since no notable differences can be detected between samples from degraded operations and those from a normally operating SEV WWTP. These normalizations incorporate the normal variability of WWTP quality and can thus be used independently of upstream conditions.

From a logical standpoint, the residual bacterial concentration after disinfection is significantly correlated (Spearman, $\alpha = 0.05$) with the C × t of PFA normalized to the initial bacterial concentration for both *E. coli* ($r = -0.604 - p$-value <0.0001) and intestinal enterococci ($r = -0.735 - p$-value <0.0001), in following a decreasing power law. The higher the initial bacterial concentration, the higher the residual concentration after disinfection for a given applied C × t. In addition, the variability in the residual concentration of intestinal enterococci is notably higher than that of *E. coli*. Using this normalized chart that takes into account influent variability, the PFA requirement in the SEV WWTP discharge to reach the quality targets of 900 and 330 MPN/100 mL can be estimated at around 1.5 ppm.min of PFA per logarithmic concentration unit of *E. coli* and around 4 ppm. min of PFA per logarithmic concentration unit of intestinal enterococci.

Since an instantaneous in-line measurement of fecal bacteria is infeasible, it would be worthwhile to use a surrogate to regulate PFA injection. Given that TSS and COD were found to be correlated with the residual bacterial concentrations, both parameters were tested during the normalization step and for both, the residual concentrations were significantly correlated with the normalized C × t of PFA applied for *E. coli* ($r = -0.690 - p$-value <0.0001 for TSS and $r = -0.643 - p$-value <0.0001 for COD) as well as intestinal enterococci ($r = -0.706 - p$-value <0.0001 for TSS and $r = -0.643 - p$-value <0.0001 for COD), in following a decreasing power law. According to these normalized charts, a normalized C × t of 1 ppm.min of PFA/mg TSS or 0.5 ppm.min of PFA/mgO$_2$-COD proves to be sufficient to ensure a reduction in *E. coli* concentration below the quality target of 900 MPN/100 mL. A normalized C × t of 2 ppm.min of PFA/mg TSS or 1 ppm. min of PFA/mgO$_2$-COD is sufficient to ensure a reduction in *E. coli* concentration below this same quality target of 900 MPN/100 mL. Since TSS can be easily estimated in-line and instantaneously by means of a turbidity probe, and COD by a 3D fluorescence probe (Goffin *et al.*, 2018), both parameters serve as good surrogates to regulate PFA injection, provided that the mathematical correlations for the SEV WWTP discharge between TSS and turbidity or between COD and 3D fluorescence are well established. A more in-depth analysis of PFA performance predictability with respect to wastewater quality parameters is performed at the industrial scale in Part 2 – Chapter 3.

1.5.2 Investigation of the role of TSS in the variability of PFA effectiveness

1.5.2.1 Description of the specific experiments performed to investigate the role of TSS

The effect of TSS alone is not directly observable by comparing PFA trials performed on different samples since the initial bacterial concentrations increase with TSS concentration in SEV WWTP discharge, as well as with other quality parameters. The TSS content in the 24 disinfection trials conducted previously is indeed significantly correlated with both *E. coli* ($r = 0.773 - p$-value $= 0.001$) and intestinal enterococci ($r = 0.637 - p$-value $= 0.006$) concentrations. It is thus impossible to directly determine if the correlation between disinfection effectiveness and TSS is due exclusively to the increase in bacterial content or if the TSS themselves have a protection effect against fecal bacteria.

To isolate the effect of TSS, specific laboratory tests have been performed in water samples spiked with different amounts of TSS collected from SEV WWTP discharge samples by means of settling in the laboratory. For each test, the TSS were collected from the same sample of water used for the disinfection tests;

these tests were carried out by spiking raw TSS (Oct 24–31, 2018, Nov 7, 2018 and Jan 30, 2019) or autoclaved TSS (Nov 7, 2019 and Feb 6–13, 2019). It was assumed that spiking autoclaved TSS would increase the TSS content yet without increasing the bacterial content, thus making it possible to isolate the effect of a TSS content modification and determine whether the TSS could protect bacteria from PFA. The tests were performed at a contact time of 10 min and a dose of 1.0 ppm of PFA (C × t of 10 ppm.min), except for the trial conducted on Jan 30, 2019, in which a lower dose of 0.5 ppm (C × t of 5 ppm.min) was applied in order to exacerbate the trends. The overall quality of the six SEV WWTP discharge samples (before TSS spiking) was normal, with average concentrations of: 13 $\pm$ 8 mg/L for TSS, 6.2 $\pm$ 0.4 mgC/L for DOC, and 26 $\pm$ 7 mgO$_2$/L for COD.

1.5.2.2 Effectiveness of PFA disinfection at increasing TSS concentration in SEV WWTP discharge

Figure 7 displays the results of these specific tests performed: (a) with raw TSS, and (b) with autoclaved TSS.

The spiking did not yield a major variation in fecal bacteria concentrations for the samples spiked with raw TSS. For these samples, the initial concentrations were quite comparable across the various samples, with a variability of less than a factor of 2 or 3, which is rather low considering the analytical method adopted for bacterial quantification. For example, *E. coli* concentrations for the Oct 24, 2018 samples were 13,800, 17,100, 10,100 and 15,700 MPN/100 mL respectively for TSS concentrations of 2, 7, 9 and 24 mg/L. Similarly, the intestinal enterococci concentrations for these same samples were 3150, 4630, 4520 and 2560 MPN/100 mL. The initial bacterial concentrations can thus be considered comparable in the samples on any given day.

Figure 7a shows two types of behavior for *E. coli*. For the Oct 24, 2018 and Oct 31, 2018 samples, no clear negative influence of TSS concentration can be observed as the residual bacterial concentration does not increase with TSS content in the sample after 10 ppm.min of PFA. For the Nov 7, 2018 and Jan 30, 2019 samples, an increase in residual bacterial concentration is observed above 20–30 mg/L of TSS. This phenomenon seems more pronounced on Jan 30, 2019, most likely because of the lower C × t of applied PFA (5 ppm.min). For intestinal enterococci, a negative effect of TSS content in the sample to be disinfected is observed, except for the Oct 24, 2018 sample. An increase in residual bacterial concentration with TSS content in the sample after 10 ppm.min of PFA is observed above 20–30 mg/L for both the Oct 31, 2018 and Jan 30, 2019 samples, and above 12–20 mg/L on Nov 7, 2018.

Regarding the trials performed by spiking autoclaved TSS, the exact same observation can be made for the effect of TSS spiking on the variation in initial bacterial concentrations, which can be considered comparable in the samples for any given day.

The results obtained with autoclaved TSS (Figure 7b) are very consistent with those from raw TSS. Both the autoclaved and raw TSS spiking tests were in fact held on Nov 7, 2018. For this particular day, no negative influence of autoclaved TSS on the residual *E. coli* concentration was observable until a concentration of 23–26 mg/L of TSS, whereas such an influence could be observed as of a concentration of 30 mg/L of TSS with raw TSS. Similarly, the negative effect of autoclaved TSS was observable on the residual concentration of intestinal enterococci from a concentration of 14–15 mg/L of TSS, while observation could only be made from a concentration of 12–20 mg/L of TSS for raw TSS. For the other tests with autoclaved TSS (i.e., Feb 6, 2019 and Feb 13, 2019), no influence of autoclaved TSS was recorded during the Feb 6, 2019 tests for either *E. coli* or intestinal enterococci, yet an increase in the residual *E. coli* concentration was detected on the Feb 13, 2019 tests as of a concentration of 14–15 mg/L of TSS.

Such results tend to confirm that TSS concentration in the wastewater to be disinfected constitutes a key parameter with a potential negative effect on PFA disinfection performance as TSS content increases even if the initial bacterial concentrations are stable. In addition, the sensitivity of intestinal enterococci to TSS increases seems to be higher than that of *E. coli*; however, the exact mechanism explaining this effect has not been clearly identified. Further experiments should thus be conducted in order to confirm this effect and identify if the TSS truly protect bacteria from PFA or if TSS consume a fraction of the PFA, thereby reducing the actual applied dose.

1.6 IMPACT OF PFA APPLICATION IN PARTIALLY TREATED WASTEWATER ON PFA DOSE REQUIREMENTS
1.6.1 Description of disinfection experiments performed in raw and settled wastewater

Several PFA disinfection trials have been performed in partially treated wastewater; the aim here was to investigate the evolution of the PFA C × t required to achieve a given removal effectiveness when PFA is applied to a highly degraded quality of wastewater. Few studies have actually reported on the disinfection effectiveness of PFA applied to settled wastewater (Gehr *et al.*, 2009) or to combined sewer overflow (Chhetri *et al.*, 2014, 2015; McFadden *et al.*, 2017; Tondera *et al.*, 2016), yet this solution is sometimes considered as an option to reduce the river's fecal bacteria inputs.

These experiments were conducted using the same laboratory scale disinfection method as previously described (Figure 1) with distinct raw wastewater and settled wastewater (physicochemical lamellar settling effluent) samples from the Seine Centre (SEC) WWTP. This WWTP treats 240,000 m^3/day of Parisian wastewater by means of pre-treatment, physicochemical lamellar settling and three-stage biofiltration that serves to eliminate carbon and nitrogen pollution. This WWTP layout is presented in detail in Rocher *et al.* (2012). The raw wastewater was also diluted with distilled water to create artificial sewer overflow samples. Four dilutions containing respectively 15, 40, 60 and 85% wastewater were applied. It is considered that samples containing 40% raw wastewater simulate the initial flow of a sewer overflow event, while 15% of the samples simulate the typical overflow (Chhetri *et al.*, 2014; Passerat *et al.*, 2011).

PFA disinfection trials were conducted on five different samples for settled wastewater (March 25, 2019, May 2, 6, 9 and 13, 2019), seven for raw wastewater and diluted raw wastewater (Dec 11, 13 and 18, 2018, May 2, 6, 9 and 13, 2019). For each trial, a contact time of 10 min was applied as well as PFA doses of 2, 6 and 10 ppm, resulting in a C × t of 20–100 ppm.min. As regards the diluted raw wastewater, three samples (Dec 11, 13 and 18, 2018) corresponded to 60 or 85% wastewater and four samples (May 2, 6, 9 and 13, 2019) corresponded to 15 or 40% wastewater. The overall quality of the seven raw wastewater samples (before dilution) was usual, with average concentrations of: 190 ± 68 mg/L for TSS, 24.9 ± 14.1 mgC/L for DOC, 356 ± 101 mgO$_2$/L for COD, 121 ± 63 mgO$_2$/L for BOD$_5$, 38.7 ± 6.1 mgN/L for TKN, and 3.8 ± 0.8 mgP/L for TP. The overall quality of the five settled SEC wastewater samples was normal, with average concentrations of: 35 ± 10 mg/L for TSS, 31.3 ± 17.3 mgC/L for DOC, 161 ± 49 mgO$_2$/L for COD, 48.7 ± 12.2 mgO$_2$/L for BOD$_5$, 37.5 ± 7.5 mgN/L for TKN, and 1.5 ± 0.8 mgP/L for TP.

1.6.2 PFA effectiveness in removing fecal bacteria in raw and settled wastewater

Figure 8 displays the results of the PFA disinfection trials performed on raw, settled and diluted raw wastewater from the SEC WWTP, with the left side results comparing raw and settled wastewater and the right-side results from diluted raw wastewater. In both cases, the median concentrations of bacteria have been plotted on top of the figure with min and max values as error bars, while the average logarithmic removals with standard deviations as error bars are plotted at the bottom.

Regarding the case of partially treated wastewater, the initial concentrations of *E. coli* and intestinal enterococci are, logically, higher than in the SEV WWTP discharge (2–3 log for *E. coli* and 1–2 log for intestinal enterococci).

For raw wastewater, a C × t of 100 ppm.min is insufficient to reduce the concentrations of both bacteria below the bathing quality limits of 330 and 900 MPN/100 mL. The increase of C × t in this type of water allows increasing the logarithmic removal from 0.96 ± 0.78 at 20 ppm.min to 3.23 ± 0.67 at 60 ppm.min and to 3.33 ± 0.81 at 100 ppm.min for *E. coli*, and from 0.68 ± 0.48 at 20 ppm.min to 3.10 ± 0.89 at 60 ppm.min and to 3.35 ± 0.85 at 100 ppm.min for intestinal enterococci.

For settled wastewater, a C × t of 100 ppm.min is insufficient to reduce the concentrations of both bacteria below the bathing quality limits of 330 and 900 MPN/100 mL. The increase of C × t in this type of water serves to increase the logarithmic removal from 2.41 ± 0.56 at 20 ppm.min to 3.10 ± 0.32 at 60 ppm. min and to 3.09 ± 0.31 at 100 ppm.min for *E. coli*, and from 0.86 ± 0.73 at 20 ppm.min to 3.46 ± 0.28 at 60 ppm.min and to 3.46 ± 0.28 at 100 ppm.min for intestinal enterococci. The logarithmic removals are thus comparable (60 and 100 ppm.min) or slightly higher (20 ppm.min) in settled wastewater compared to raw wastewater, however the residual bacterial concentrations are lower thanks to lower initial concentrations. A C × t above 100 ppm.min is thus required to efficiently disinfect raw or settled wastewater, in comparison with a C × t of 10–20 ppm.min for the SEV WWTP discharge.

To the best of the author's knowledge, no results are available in the literature on PFA effectiveness on raw wastewater. In contrast, similar experiments were performed by Gehr *et al.* (2009) on physicochemical treatment effluent (10–100 mg/L of TSS and 37–238 mgO$_2$/L of COD) at both the batch and pilot scales. At the batch scale for a contact time of 10 min, they obtained fecal coliform removals of approximately 3 log at 5 ppm.min of PFA, slightly above 3 log at 20 ppm.min and up to 6 log at 40 ppm.min. At the pilot scale for a contact time of 45 min, average fecal coliform removals of 0.5 log at 45–90 ppm.min, 2.5 log at 90–135 ppm.min, 3 log at 135–180 ppm.min and up to 5.5 log at 225–270 ppm.min were recorded with PFA. For intestinal enterococci, removals of 4–6 log were observed for a contact time of 45 min at 225–270 ppm.min of PFA.

As regards the case of artificial combined sewer overflows (CSOs), the initial concentrations of both *E. coli* and intestinal enterococci decreased with the dilution of raw wastewater, with a maximum difference of around 1 log between the concentrations in 100 and 15% raw wastewater.

For *E. coli*, the dilution rate increase led to a decrease in residual concentration after 20 ppm.min of PFA, with median concentrations of: 3,332,461, 1,792,500, 18,863, 2695 and 2598 MPN/100 mL respectively for 100, 85, 60, 40 and 15% raw wastewater. However, at 60 and 100 ppm.min, the residual bacterial concentrations were similar among the various dilution rates, that is, around 1000–3000 MPN/100 mL. These values were correlated with an increase in the logarithmic removal rate between 100 and 85% raw wastewater and more heavily diluted wastewater for a C × t of 20 ppm.min, which rose from around 1 log to just below 3 log, in contrast with 60 and 100 ppm.min, at which logarithmic removal remained stable between 3 and 4.5 log.

For intestinal enterococci, the decrease in residual concentration with an increasing dilution rate was observable for all three C × t. For example, the median concentrations after 100 ppm.min of PFA were 1495, 391, 107, 54 and 37 MPN/100 mL respectively for 100, 85, 60, 40 and 15% raw wastewater. As was the case for *E. coli*, the logarithmic removal of intestinal enterococci increased with a dilution rate at 20 ppm.min of PFA, from below 1–3 log, but remained stable at 60 and 100 ppm.min between 3 and 4 log.

Consequently, for both bacteria and except for a low C × t of 20 ppm.min, the logarithmic removal rate was relatively stable for the various dilution rates, and improved residual bacterial concentrations resulted from the initial decrease in concentration due to dilution.

Several papers have studied the disinfection of CSO by PFA (Chhetri *et al.*, 2014, 2015; McFadden *et al.*, 2017; Tondera *et al.*, 2016). In particular, Chhetri *et al.* (2014) performed laboratory scale disinfection tests on diluted raw wastewater with 5, 15 and 40% raw wastewater; they obtained logarithmic removals comparable to those in diluted raw SEC wastewater. The removals of intestinal enterococci were 1 and 3.5 log, respectively, for 40 and 80 ppm.min in 40% raw wastewater, which corresponds to the first flush. For 5% raw wastewater, corresponding to the maximum of an overflow event, removals were higher, i.e. respectively 4 and 3.5 log for *E. coli* and intestinal enterococci at 20 ppm.min of PFA or up to 5 log at 80 ppm.min. Tondera *et al.* (2016) studied the application of PFA at an industrial scale on CSO at a significantly higher C × t compared to this study, with doses between 12 and 24 ppm and a maximum contact time of 30 min. They obtained bacteria reductions of 100–1000 CFU/100 mL for both *E. coli* (initial concentration: 10^6 CFU/100 mL) and intestinal enterococci (initial concentration: 10^6 CFU/100 mL). The associated logarithmic removals were: 3.1 ± 1.7 log for *E. coli*, and 2.6 ± 1.5 log for intestinal enterococci. Comparable removals were reported by Chhetri *et al.* (2015) at the industrial scale with comparable initial concentrations, lower PFA doses between 4 and 8 ppm and a contact time of 20 min (C × t between 80 and 160 ppm.min). These authors also recommended regulating the PFA dose injection based on water quality and then treating the first 60 min of an event with a higher PFA dose than the rest of the overflow.

In terms of the PFA C × t required to achieve a given disinfection effectiveness, a C × t of 100 ppm.min was insufficient to reduce the median concentration of *E. coli* below the bathing quality limit of 900 MPN/100 mL for any type of diluted raw wastewater. However, residual concentrations after 100 ppm.min of PFA varied between 38 and 3050 MPN/100 mL for 15% raw wastewater in two of the four tests where concentrations were below 900 MPN/100 mL after disinfection. In contrast, a C × t between 60 and 100 ppm.min was sufficient to reduce the median concentration of intestinal enterococci below the bathing quality limit of 330 MPN/100 mL for diluted wastewater containing 60% or less raw wastewater. For a 15% raw wastewater level, a C × t of 20 ppm.min seems to allow reducing the median intestinal enterococci concentration below 330 MPN/100 mL. It can thus be assumed that a PFA C × t of approximately 100 ppm.min would be required to disinfect Parisian CSO, hence 5–10 times higher than for SEV WWTP discharge. This finding means that in the case of in-line injection of PFA with very limited contact times, that is, between 1 and 5 min, and in considering the hypothesis that concentration and time have the same effect on disinfection effectiveness, the required PFA dose would be 20–100 ppm, which represents a significant input of carbon into the river, theoretically 15.8–79 mgC/L respectively (1 ppm of PFA injected injects 0.79 ppm of C). To limit this carbon input to less than 1–2 mgC/L, as in the case of the SEV WWTP discharge, dedicated high-volume contact tanks ensuring a contact time of 50–100 min would need to be built.

Along with a notably higher required PFA C × t, thus requiring long contact times or the discharge of large quantities of carbon into the river, the disinfection of non-biologically treated wastewater with PFA could result in a greater risk of byproduct formation since raw or settled wastewater and CSO possess a higher quantity of organic matter and particles. Chapters 3 and 4 of Part 1 provide more information on this topic.

1.7 PFA EFFECTIVENESS IN REMOVING OTHER MICROORGANISMS

In addition to *E. coli* and intestinal enterococci, F-specific bacteriophages and SSR (sulfite reducing bacteria) spore analyses were performed for five PFA disinfection tests (Sept 11–18, 2018, Jan 30, 2019 and Feb 6–12, 2019) to generate information on PFA effectiveness regarding other pathogens. As a reminder, these tests were conducted with PFA doses of between 0.8 and 2 ppm and a 10-min contact time, in using SEV WWTP discharge samples. Both parameters were analyzed according to Standards ISO 10705-3 and NF EN ISO 10705-1 for F-specific bacteriophages, and NF EN 26461-2 for SSR spores. The concentrations before and after disinfection, as well as logarithmic removals, are listed in Table 1.

SSR spores were detected in the seven SEV WWTP discharge samples, with a median concentration of 5000 CFU/100 mL, while F-specific bacteriophages were only detected in two of the five samples considered, with a concentration close to the LOQ of 30 CFU/100 mL. The SEV WWTP operations (whether normal or degraded) had no impact on the initial concentrations of these pathogens. These concentrations were in the range of the data measured on SEV WWTP discharge by the SIAAP Authority over the period 2014–2017, with median ± min-max values of 860 ± 4–2700 UFC/100 mL for SSR spores ($n = 27$ values) and 150 ± 14–7800 UFC/100 mL for F-specific bacteriophages ($n = 27$ values), even though they lie in the top of the range for SSR spores and the bottom of the range for F-specific bacteriophages.

PFA at a limited C × t of 8–20 ppm.min seems to exert no or only a very limited effect on SSR spores for five of the seven tests, with logarithmic removals lying below 0.5, but better removals were achieved under these same conditions for the other two tests, that is, between 1 and 4 log. The effect of PFA on F-specific bacteriophages could not be determined in light of the very small initial concentrations close to the LOQ. The effect of PFA on pathogens other than

E. coli and intestinal enterococci in the SEV WWTP discharge is thus uncertain, and complementary experiments should be performed.

In the literature, information on PFA is limited although several papers deal with the effectiveness of peracids on pathogens in wastewater (Gehr *et al.*, 2009; Karpova *et al.*, 2013; Luukkonen & Pehkonen, 2017; Mora *et al.*, 2018; Tondera *et al.*, 2016). In particular, Luukkonen and Pehkonen (2017) conducted a critical review of PFA disinfection.

As regards bacteria, most authors have indicated the good effectiveness of PFA. For example, Tondera *et al.* (2016) reported removals of fecal coliforms *Aeromonas* spp. and *C. perfringens* of 1.8 to 3.1 log in CSO with a dose of 12–24 ppm of PFA and a contact time of 30 min (C × t of 360–720 ppm.min). The good effectiveness of PFA on fecal coliforms has also been reported in most publications (Gehr *et al.*, 2009; Karpova *et al.*, 2013; Luukkonen & Pehkonen, 2017; Ragazzo *et al.*, 2013). In addition, Karpova *et al.* (2013) noted a logarithmic removal of *Salmonella* spp. equal to 3 log for a C × t of 20 ppm.min.

Regarding spores and cysts, e.g. *Clostridia* spp. or *Giardia* spp., they have been identified as more recalcitrant to PFA (Karpova *et al.*, 2013) or PAA (Luukkonen & Pehkonen, 2017). A PFA C × t of 10 ppm.min applied in a WWTP discharge does indeed lead to a limited removal of below 1 log for *Clostridia* spp. and *Giardia* spp. (Karpova *et al.*, 2013). In CSO, Tondera *et al.* (2016) reported no significant removal of *Giardia lamblia* despite a high C × t of 360–720 ppm.min. Gehr *et al.* (2009) found *Clostridia* removals of 1–2 log at PFA C × t of 450–540 ppm.min in physico-chemically treated wastewater effluent. Overall, the disinfection effectiveness of PFA is higher than PAA with respect to *Clostridia* (Mora *et al.*, 2018).

As for bacteriophages, MS2 and DNA bacteriophages have been reported as more sensitive to PFA than intestinal enterococci and *E. coli* (Karpova *et al.*, 2013). In fact, these authors reported MS2 and DNA bacteriophage removals of 1 log for a limited PFA C × t of 17–22 ppm.min. Gehr *et al.* (2009) reported F-specific bacteriophage removals of 1–2 log at a PFA C × t of 450–540 ppm. min in a physico-chemically treated wastewater effluent. Logarithmic removals of somatic coliphages equal to 2.7 ± 1.6 in CSO were observed by Tondera *et al.* (2016) with a very high PFA C × t of 360–720 ppm.min. In a WWTP discharge, Karpova *et al.* (2013) observed higher removals of somatic coliphages of 4 log for a significantly lower C × t of 10 ppm.min.

Regarding human viruses like adenovirus, polyomavirus, norovirus, rotavirus and enterovirus, no results are available for PFA, but PAA has been reported as inefficient. Tondera *et al.* (2016) concluded that viruses are not inactivated by PAA, even at a high C × t of 120–240 ppm.min. However, Luukkonen and Pehkonen (2017) indicated that the virus inactivation mechanism by PAA may occur by damaging the virus surface, for example, the protein coat of the sites needed to infect the host cells. Such information has not been reported for PFA.

These elements therefore suggest that in the case of viruses, spores, cysts or bacteriophages being present in WWTP discharge, a higher PFA C × t would be required to remove them in comparison with bacteria such as *E. coli* or intestinal enterococci.

Key Points

- In the SEV WWTP discharge, the required PFA C × t to achieve a given disinfection rate is 2–4 times less than PAA, thus leading to a significantly lower potential carbon input into the river.
- Laboratory scale PFA disinfection trials performed with the SEV WWTP discharge indicate that a PFA C × t of 10–20 ppm.min is sufficient to ensure a reduction in *E. coli* and intestinal enterococci concentrations to below 900 and 330 MPN/100 mL, respectively. Intestinal enterococci are more resistant to PFA in wastewater than *E. coli*.
- The residual concentrations of *E. coli* and intestinal enterococci after a given C × t of PFA disinfection are correlated with not only the initial bacterial concentrations, but also TSS and COD, which serve as good surrogates to regulate the PFA dose injection.
- TSS content has an impact on the PFA disinfection effectiveness by means of both an increase in the initial bacterial concentration and an intrinsic effect above 20 mg/L of TSS.
- PFA is efficient in removing fecal bacteria from Parisian combined sewer overflows or partially treated wastewater, yet the required C × t is at or above 100 ppm.min. This finding would lead to applying a high PFA dosage, which then implies high carbon inputs into the river or the necessity of long contact times, in addition to higher risks in terms of organic matter evolution and byproduct formation.
- Limited information is available regarding the effect of PFA on other pathogens, but the scarce literature available and the results obtained in the SEV WWTP discharge indicate a limited effect at a C × t of 10 20 ppm.min on spores, cysts, bacteriophages and human viruses. This issue should however be investigated more thoroughly.

Chapter 2

Fecal bacteria regrowth and viability after disinfection with PFA

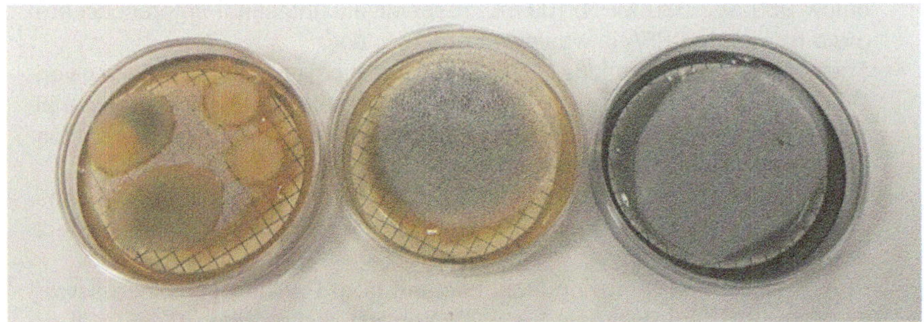

© Thierry Pigot, UPPA

2.1 INTRODUCTION

The bactericidal efficacy of PFA against fecal contamination control germs is now well documented and has been demonstrated in this study (see Chapter 1). However, when using a tertiary disinfection treatment, it is important to evaluate the regrowth potential of the germs contained in the treated water. This is especially critical when wastewater reuse applications are being considered. Bacterial inactivation can in fact lead to the production of viable non-cultivable bacteria as an initial response, and various repair processes can then renew the ability to reproduce after a certain length of time (usually several hours). This phenomenon has been studied with well-known disinfection processes and used at the industrial scale for several years (UV-C, chlorination, ozone) or with advanced oxidation processes (e.g., photocatalysis, UV/H_2O_2).

doi: 10.2166/9781789062106_0022

The phenomenon of cellular repair has been demonstrated in several studies during the UV-C disinfection of wastewater (Bohrerova & Linden, 2007; Bohrerova Zuzana *et al.*, 2015). Such a disinfection process seems to depend on the UV dose applied and moreover appears to be reached 24 hours after irradiation (Oguma *et al.*, 2002). Several repair mechanisms have been identified, some in the dark while others photoinitiated (Shang *et al.*, 2009).

Similarly, the simple ozonolysis of a secondary effluent leads to the regrowth of *E. coli* and coliforms (Malvestiti & Dantas, 2018). The presence of hydroxyl radicals (generated from coupling O_3/H_2O_2 or UV-C/H_2O_2) seems to mitigate this phenomenon. On the other hand, the presence of OH inhibitors (nitrates, carbonates) would seem to favor it. Some studies mention a synergistic effect when ozone is introduced in combination with photocatalysis (Mecha *et al.*, 2017): while ozone and photocatalysis used separately lead to bacterial regrowth, their combined use seems to prevent it. Tertiary treatment by photocatalysis alone (TiO_2/UV-A) does not prevent regrowth on various heterotrophic microorganisms, including *E. coli* and enterococci (Biancullo *et al.*, 2019). Several studies suggest that bacterial regrowth is slower during H_2O_2/sun disinfection than with chlorination (Fiorentino *et al.*, 2015; Giannakis *et al.*, 2015; Li *et al.*, 2013).

The chemical disinfection of secondary effluents is another widely employed process. In the case of chlorination, bacterial regrowth can occur, especially when low doses are applied (Li *et al.*, 2013), and is heavily dependent on the microorganisms under study. In the case of disinfection with peracids, recent results with peracetic acid (PAA) have been published; Zhang *et al.* (2019b) determined the minimum dose of PAA to prevent the regrowth of planktonic cells. This dose is dependent on the amount of dissolved organic matter (DOM) present in the wastewater. A comparison with chlorination shows that for this microorganism, chlorination requires lower doses yet is much more sensitive to the presence of DOM. Zhang *et al.* (2019a) also showed that while both PAA and chlorine prevented bacterial regrowth, they were inactive against a plasmid incorporated into the DNA of bacteria. To the best of our knowledge, no data have been published with PFA concerning these investigations; consequently, the objectives of this chapter are twofold: (1) to study the potential for regrowth of fecal contamination control germs in a performic acid-treated environment; and (2) to provide information on the effect of performic acid on these germs in order to explain the effect being observed.

2.2 EXPERIMENTAL DESCRIPTION
2.2.1 Fecal bacteria regrowth after PFA disinfection

The experimental procedure used to study bacterial regrowth is summarized in Figure 9 (Steps 1 and 2.1); it has been adapted from previous studies (Fiorentino *et al.*, 2015; Zhou *et al.*, 2017). In Step 1, two bacterial inocula (*E. coli* and intestinal enterococci) were prepared from SEV WWTP discharge water using restrictive agar media. Once these strains had been isolated, they were stored at $-20°C$ in a mixture of 30% glycerol for 70% culture broth.

The bacteria were first cultured from agar and then from a liquid medium once the strain had developed sufficiently, resulting in a solution in a decreasing exponential phase whose bacterial concentration could be estimated by an absorbance measurement at 600 nm (0.1 absorbance unit $= 10^7$ bacteria/mL). The bacterial suspension was then diluted to a realistic bacterial concentration compared to the usual concentrations of the SEV discharge (Rocher & Azimi, 2016): *E. coli* between 10^5 and 10^6 MPN/100 mL and intestinal enterococci between 10^4 and 10^5 MPN/100 mL.

This step of the experimental procedure allows working at controlled concentration of bacteria composed of pure strains but coming from real SEV discharge samples.

The kinetic monitoring of bacterial regrowth was performed by either microplate or agar plate enumeration. Agar analyses were preferred on samples treated with PFA since this method has a lower quantification level (LOQ <3 NPP/100 mL). Microplates were used for untreated control samples with bacterial concentrations greater than 10^3 NPP/100 mL. The initial point was established by both methods in order to verify that they yielded equivalent results.

In Step 2.1, a bacterial inoculum containing *E. coli* and intestinal enterococci was introduced into autoclaved discharge water in order to achieve realistic concentrations. Some of this bacterial suspension was treated with PFA at a treatment rate of 0.8 ppm for 10 min (C $\times$ t of 8 ppm.min) and another part was used as a control. The sample treated with PFA was then mixed with autoclaved Seine River water (90/10 volume mixture) to best simulate actual mixing of the SEV discharge and Seine River water at the WWTP discharge point (unfavorable case of the Seine at a low water flow rate). A cultivable bacteria count according to standard methods was carried out at different times after disinfection (up to 24 hours) on both samples.

Two experimental conditions were applied during the tests as follows:

- The first considered a regrowth at room temperature under laboratory light at constant temperature (26°C). This experiment was carried out in triplicate.
- The second condition considered a regrowth under irradiation simulating the solar spectrum (visible lamps + UV-A lamps) at room temperature (26°C), with the experiment being performed in duplicate.

These experiments were carried out in a temperature-controlled chamber equipped with four fluorescent tubes (either three tubes emitting in the visible light and one tube emitting in UV-A, or two tubes emitting in the visible spectrum and two emitting in UV-A). The distance of the tubes from the sample was adjusted such that the light power was comparable to that of sunlight (i.e., between 4 and

$6\ \text{mW/cm}^{-2}$ in the UV-A range). Measurements were controlled by a spectroradiometer (CAS 120, Instrument Systems). In the two experiments, the UV-A light power was 4 and 8 mW/cm^2.

2.2.2 Viability tests

Two types of analyses were conducted for this part of the study: (1) bacterial counting of samples before disinfection by the methods described above; and (2) determination of bacterial viability before and after disinfection by cytometry (cytometric counting).

For the tests carried out on untreated samples, both methods produced comparable results (at uncertainties close to those of the method, i.e., 20%). Only untreated samples were monitored by both methods.

The count performed by flow cytometry was performed on a BD Accuri C6 device (BD Biosciences – Becton Dickinson, France SAS, Grenoble, France); this apparatus is equipped with a scatter detector and three fluorescence detectors (WL1, WL2 and WL3).

The viability analyses were carried out using the BD Cell viability kit provided by BD Biosciences. This kit contains various solutions, namely:

- a 'rainbow balls' solution to ensure a response from the device's various detectors;
- a calibration bead solution for a quantitative determination of the analyzed objects;
- an orange thiazole (OT) solution (500 μL at a concentration of 42 μmol/L in DMSO); this compound responds to both WL1 and WL2 detectors and colors all cells;
- a solution of propidium iodide (PI) (500 μL at a concentration of 4.3 mmol/L in water); this compound responds to WL3 and colors the lysed cells.

A typical experiment necessitates several control cytometric analyses:

(1)   500 μL of physiological water + rainbow beads (instrument validation);
(2)   500 μL of physiological water (matrix validation);
(3)   500 μL of physiological water + calibration beads (validation quantification);
(4)   500 μL of bacterial suspension (concentration around 10^5 bacteria/mL) without dyes (sample control);
(5)   500 μL of bacterial suspension (around 10^5 bacteria/mL) with fluorescent PI and TO dyes (marking control, qualitative analysis);
(6)   500 μL of bacterial suspension (around 10^5 bacteria/mL) with PI+TO fluorescent dyes + calibration beads (sample measurement, qualitative and quantitative analyses);
(7)   500 μL of bacterial suspension + NaN_3 + PI+TO fluorescent dye (lysed bacterial control).

After these controls, the sample analysis by cytometry could be achieved by adding 5 µL of each dye and 50 µL of calibration beads to 500 µL of each sample (bacterial suspension treated with 0.8 ppm performic acid and an untreated suspension). The BD Accuri™ C6 Plus software interface provided quick access to the collection, analysis and statistics functions. The raw data could also be analyzed using the R program 'rattle package' (Williams, 2011).

The interpretation of results is obtained by plotting WL1 and WL3 fluorescence intensities on a 2D graph. Given the emission properties of the dyes, the best distinction between populations is derived by monitoring the fluorescence intensities on WL1 and WL3:

- a high intensity on WL1 and a low intensity on WL3 corresponds to viable bacteria;
- an increase in the WL3 intensity corresponds to bacteria that have already been lysed.

2.3 RESULTS
2.3.1 Regrowth kinetics

The regrowth results, under dark conditions of samples reconstituted from real autoclaved matrices doped with isolated *E. coli* (high) and intestinal enterococci (low) germs and then disinfected with 0.8 ppm PFA, are shown in Figure 10. This experiment was carried out in triplicate with different initial matrices and identical pure strains. At the initial time ($t = 0$), the three histograms correspond to the initial concentrations of fecal bacteria in the non-disinfected control samples. At the other times ($t = 1$, 3, 6, 8 and 24 h), the three histograms on the left part of the figure correspond to the non-disinfected control samples, while the three histograms on the right part of the figure correspond to these same samples yet disinfected with PFA (8 ppm.min). Each color thus corresponds to the same experiment.

For *E. coli* and intestinal enterococci, the concentrations of the three untreated controls (July 17, 2018, July 10, 2018 and Sept 25, 2018) increase over 24 hours, mostly after 3 hours. This result indicates that the conditions applied during the experiments are indeed favorable to bacterial growth in the controls. This phenomenon is less obvious for intestinal enterococci.

Adding 0.8 ppm PFA induces results systematically lower than LOQ: no cultivable bacteria are quantified even after 24 hours, which proves the absence of regrowth. It should be noted that the treatment efficacy on *E. coli* tends to be quite good, as well observed in the present case. Studies conducted at other sites have yielded comparable results (Ragazzo *et al.*, 2013).

For intestinal enterococci, the treatment effectiveness of 0.8 ppm PFA is lower: for both the July 10, 2018 and Sept 25, 2018 experiments, 70 and 80 CFU/100 mL were quantified after disinfection, respectively; however, no increase over 24 hours was observed experimentally. The concentration does tend to decrease slightly for the July 10, 2018 experiment (from 70 to 35 between 1 and 24 h) and remain stable for the Sept 25, 2018 experiment (80 CFU/100 mL). The July 17, 2018 experiment results are similar to *E. coli*, i.e. below the LOQ until 24 hours. The conclusion is thus the same as for *E. coli*: no regrowth of enterococci after PFA treatment even though the disinfection efficacy is lower, as previously demonstrated in Chapter 1 and by other studies (Ragazzo *et al.*, 2013). To the best of our knowledge, these are the first results describing regrowth experiments involving the PFA disinfection of wastewater. Some of the recent results published for PAA disinfection have shown inhibited regrowth of planktonic and biofilm bacteria after disinfection (Zhang *et al.*, 2019a).

These same experiments were repeated with simulated solar irradiation. Sunlight was simulated by a set of lamps emitting in both the visible and the UV-A wavelength. Several studies have revealed that light can stimulate cell repair processes after disinfection treatment and moreover that bacterial regrowth can then be observed (Giannakis *et al.*, 2016).

The UV-irradiated bacterial regrowth results of samples reconstituted from real autoclaved matrices doped with isolated *E. coli* (high) and intestinal enterococci (low) germs, and then disinfected with 0.8 ppm PFA, are shown in Figure 11.

This experiment was carried out in duplicate with different initial matrices and identical pure strains. At the initial time (i.e., $t = 0$), the two histograms correspond to the initial concentrations of fecal bacteria in the non-disinfected control samples. At the other times ($t = 2$, 4 and 6 h), the two histograms on the left part of the figure still correspond to the non-disinfected controls, while the two histograms on the right part correspond to the same samples but disinfected with PFA (8 ppm.min). Each color thus corresponds to the same experiment. Note that for one experiment (Feb 10, 2018, gray), the UV-A light power is $8 \text{ mW} \cdot \text{cm}^{-2}$ while for the other the UV-A light power is $4 \text{ mW} \cdot \text{cm}^{-2}$ (Oct 16, 2018, black).

UV-A exerts a major short-term effect (<2 h) on *E. coli* (total disinfection). The simulated solar radiation had a sublethal effect on *E. coli* populations over a brief time exposure by causing a loss of culturability and the formation of viable yet nonculturable cells (Muela *et al.*, 2000). The presence of colored DOM in the water matrix may also favor the formation under irradiation of Reactive Oxygen Species (e.g., singlet oxygen, superoxide anion), whose bactericidal properties are well known (Häder *et al.*, 2015; Maraccini *et al.*, 2016).

The effects of light on enterococci are less apparent, but they display greater resistance to UV-A irradiation. A latency period of 4 hours seems to be required for the UV-A power of 8 mW $\cdot$ cm^{-2}, while a continuous decrease is observed for irradiation at 4 mW $\cdot$ cm^{-2}.

After adding PFA, the bacterial concentration remains below the limit of quantification after 4 hours of irradiation. The increase in concentration after 2 hours in one experiment remains unexplained to this day and could be due to an external contamination. In any case, regrowth was never observed in this study either in the dark or under irradiation.

2.3.2 Fecal bacteria viability after disinfection with PFA

The objective of these experiments has been to provide additional information regarding the actual effect of PFA on fecal bacteria and, in particular, to better understand their viability after disinfection. The testing method employed, that is, flow cytometry, is applicable to monitoring the effect of a biocidal treatment or comparing various biocidal treatments (Whitton *et al.*, 2018).

Figure 12 presents the results obtained on a suspension of *E. coli* germs treated with 0.8 ppm PFA for 10 min; the initial concentration of *E. coli* was approximately 10^5 NPP/100 mL. Figure 12 shows the flow cytometry maps derived in the single test performed, both before and after disinfection, while Figure 13 provides these same results as percentages of detected bacteria, which were classified as dead, alive or damaged by the methodology used.

In Figure 12, the illustrations before disinfection (left) show the bacterial population of *E. coli* before treatment. As anticipated for a population composed mainly of living bacteria, high fluorescence intensity in WL1 and low fluorescence intensity in WL3 are observed. After treatment with 0.8 ppm PFA (right part of the figure), a low fluorescence intensity in WL1 and a high intensity in WL3 can be noted, which means that the *E. coli* are mostly dead (99.3% dead, see Figure 13). Nearly identical results have been obtained from a suspension of enterococci (99.6% dead bacteria after PFA treatment).

Figures 12 and 13 reveal that PFA causes irreversible damage to fecal contamination indicator bacteria cells and moreover that such cells are not in a 'viable non-cultivable' (VNC) state after disinfection. It does appear that the entire population of *E. coli* and intestinal enterococci are lysed after disinfection at 8 ppm.min PFA, whereas they were for the most part alive or damaged before disinfection. These results tend to agree with the regrowth experiments previously described in this chapter and those obtained with PAA, where no regrowth phenomena were observed with these bacteria (Antonelli *et al.*, 2013). In the case of PAA, several observations recorded by electron microscopy indicate that the cells are damaged and holes appear in the cell centers (Zhang *et al.*, 2019b); these results are attributed to reactions between PAA and cellular components. As for PFA, its ability to oxidize disulfide links (S-S) in cysteic acid RSO$_3^-$ and the sulfide function of methionine to methionine sulfone in protein residues may help explain its irreversible effect on bacteria (Voet & Voet, 1995).

Key points

- PFA allows for stable disinfection of wastewater, with no regrowth after 24 hours without irradiation, even though the *E. coli* concentration in the control samples increased.
- UV irradiation, which had been identified in the literature as a parameter that could mitigate the effects of disinfectants by stimulating cell repair, did not lead to regrowth after 6 hours; moreover, it even enhanced the disinfection effects.
- The flow cytometry analysis of a bacterial suspension containing either *E. coli* or enterococci shows that the cells appear to be irreversibly inactivated.
- These results, combined with those from studies conducted with PAA, suggest that PFA reacts irreversibly on cell membrane components, most likely by means of chemical oxidation on chemical functions present in proteins (disulfides, sulfides) and unsaturated fatty acids ($C = C$ double bond).

Chapter 3

Impact of PFA on organic matter and post-injection consequences

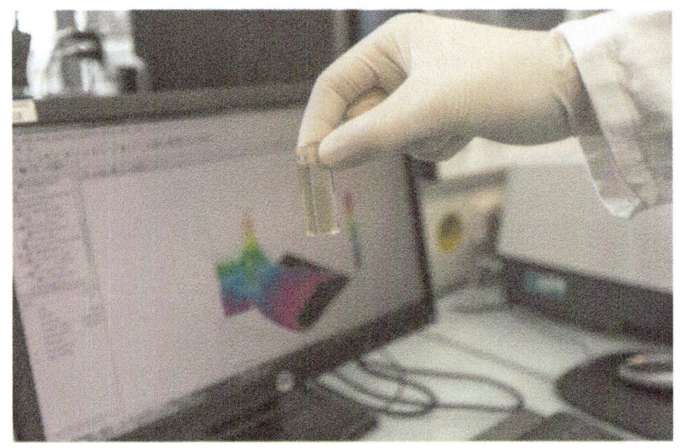

© SIAAP

3.1 INTRODUCTION

Performic acid (PFA) is an organic oxidant, which consequently is capable of affecting organic matter in wastewater. The injection of an organic compound will indeed increase the dissolved organic carbon of wastewater. This DOC addition constitutes a mix of PFA and formic acid, and the PFA instability will result in a PFA degradation to the formic acid once injected into the wastewater. The kinetics of this degradation process in wastewater and Seine River water needs to be investigated, along with the PFA degradation products, in order to

doi: 10.2166/9781789062106_0030

determine whether the environmental fate of PFA after injection is an issue. In addition, the oxidative power of PFA could cause changes in the organic matter nature, and this remains an outstanding question.

This chapter aims to present the results obtained relative to the PFA impact on SEV WWTP discharged organic matter and its fate in both wastewater and Seine River water after treatment. First, the DOC increase caused by PFA injection will be quantified and compared with the theoretical input of the PFA mixture. Next, the effect of PFA on wastewater organic matter nature will be evaluated based on 3D fluorescence. Lastly, the fate of PFA after injection will be investigated both quantitatively and qualitatively. PFA degradation kinetics are determined in both SEV WWTP discharge and Seine River water, as are the PFA half-lives. PFA degradation products can then be measured by means of liquid chromatography with conductive detection and UV detection so as to verify that only formic acid and H_2O_2 are formed.

3.2 IMPACT OF PFA ON THE SOLUBLE ORGANIC MATRIX
3.2.1 Increase in dissolved organic carbon concentration

The injection of PFA into a WWTP discharge leads to injecting dissolved organic carbon (DOC). In theory, the DOC content of the prepared PFA solution equals 0.79 ppm C/ppm PFA injected (0.19 from PFA and 0.60 from the residual formic acid), as described in Part 1, Chapter 1. To verify this assertion, DOC values from the 24 laboratory scale PFA disinfection trials performed in Part 1, Chapter 1 on SEV WWTP discharge samples, are plotted in Figure 14, which displays the average DOC concentrations measured and the standard deviations as error bars for each C $\times$ t, as well as the average values and standard deviations for the theoretical DOC concentrations. These theoretical concentrations are obtained by adding the 0.79 ppm C/ppm PFA injected to the initial DOC values.

The DOC concentrations measured after disinfection are very consistent with those obtained theoretically by adding 0.79 ppm C/ppm PFA injected to the initial DOC concentrations. Both series are indeed significantly comparable (Mann-Whitney, p-value $= 0.8397$). In addition, the average increase in calculated DOC based on actual measurements equals 0.78 ppm C/ppm PFA injected. The slight observable difference between actual measurements and theoretical values in Figure 14 can be explained by the variability of the DOC analytical method. It can thus be confirmed that injecting PFA preparation to achieve a PFA dose between 1 and 2 ppm would result in a limited increase of 0.79 and 1.58 mgC/L of the DOC.

By comparison, applying a dose of 8 ppm of peracetic acid (PAA) in SEV WWTP discharge, which is the dose required to obtain the same bacteria reduction as 2 ppm of PFA (Part 1, Chapter 1), would theoretically result in a minimum DOC increase of 2.53 mgC/L since 1 ppm of PAA contains 0.32 mgC/L and provided the PAA preparation does not contain any other organic compounds. Similarly, since high doses of 10 ppm of PFA or more are required to disinfect partially-treated Parisian wastewater or combined sewer overflows with a contact time of 10 min (Part 1, Chapter 1), the DOC increase in such a case would be very high, that is, 7.9 mgC/L or more.

3.2.2 Modifications to dissolved organic matter quality, as monitored by 3D fluorescence spectroscopy

3.2.2.1 Experimental description

The analytical protocol described below has been applied to the PFA addition tests performed on the following dates: Sept 4, 2018, Sept 11, 2018, Sept 18, 2018, Oct 30, 2018, and Nov 6, 2018. Before analysis by 3D fluorescence spectrometry, the selected samples from SEV WWTP discharge water (both with and without PFA) were filtered at 0.45 µm (GF/F, glass fiber, Whatman®). Fluorescence spectrometry measurements were performed using a 3D spectrofluorimeter (Jasco FP-8300, Japan) equipped with a 150-W xenon lamp as the excitation source. This device is controlled by the Spectra Manager II software. Measurements were conducted in a 1-cm quartz cell at a controlled temperature of 20°C in order to avoid thermal fluorescence extinction (Watras *et al.*, 2011). Fluorescence spectra (excitation-emission matrix: EEM) were measured for excitation wavelengths (Ex) ranging from 240 to 550 nm every 5 nm and for an emission wavelength (Em) ranging from 260 to 700 nm with a measurement step of 2 nm. The widths of the emission and excitation slots were set to 5 nm, with a scanning rate of 1000 nm/min.

To avoid internal filter effects due to highly absorbent species present in the samples, the species were diluted with ultrapure water to obtain an absorbance of $0.05 \, \text{cm}^{-1}$ at 254 nm before fluorescence analysis (Alberts & Takács, 2004; Lakowicz, 2010). The absorbance at 254 nm was measured in a standard 1.0-cm quartz cuvette using a UV-Vis spectrophotometer (UviLine 9400, Seconam – Xylem S.A.S., White Plains, USA). The EEM measurements (both blank and sample) were normalized by the area of the Raman peak of ultrapure water at 350 nm excitation (measured each day of analysis), thus allowing fluorescence data to be obtained in Raman units (RU) (Determann *et al.*, 1998; Lawaetz & Stedmon, 2009). Moreover, an EEM of ultrapure water (blank) was subtracted from the sample EEMs measured on the same day. The resulting EEMs were plotted using Matlab® R2013b software (Mathworks, Natick MA, USA). The 'peak-picking' fluorescence data processing protocol was applied in this study since the number of samples was not large enough to perform a mathematical decomposition of EEM into its fluorescent components (parallel factor analysis). The 'peak-picking' method consists of extracting a maximum value of fluorescence intensity observed within a delimited area of the fluorescence spectrum. Each zone delimitations and related chemical compounds from dissolved organic matter (DOM) have been compiled in Table 2 and illustrated in Figure 15.

3.2.2.2 Results

3D EEM fluorescence spectra. 3D fluorescence spectrometric analyses were performed for SEV WWTP discharge samples with the following PFA concentrations: 0, 8, 10, 12, 20, 300, and 1000 ppm.min. An example of the 3D EEM spectra obtained for two disinfection test campaigns is shown in Figure 16.

An initial approach to qualitatively interpreting 3D fluorescence spectra consists of locating fluorescence maxima with the 'peak-picking' method, as described above. For all spectra shown of SEV WWTP discharge, whether or not exposed to a PFA concentration, a similar fluorescence signature is apparently observed. The highest fluorescence intensities recorded are related to the 'humic substances-like' fluorescent component of DOM, which is known to be more resistant to biodegradation (Parlanti *et al.*, 2000) and located in the excitation-emission wavelength region called the α band (Ex/Em 330–370/420–480 nm). Other areas of interest are also observed, with lower fluorescence intensities: peak α' (Ex/Em 230–260/380–480 nm) associated with 'humic substances-like' fluorescent components mixed with a recent DOM material and peak δ related to the 'protein-tryptophan-like' component of fluorescent DOM.

Analysis of SEV WWTP discharge EEM spectra according to the operationally applicable PFA processing rates (0–20 ppm.min). Since DOM fluorescence quality and quantity indicate very similar fluorescence intensities among the three disinfection testing campaigns carried out (results not shown, relative standard deviation less than 4%), only the average results will be discussed herein. By comparing the average fluorescence intensities (Figure 17a) obtained for both the non-disinfected and disinfected SEV WWTP discharges at various PFA concentrations (operating rates: 0–20 ppm.min), a similar amount of fluorescence intensity has been observed (relative standard deviation: 3% for all peaks). It should be noted though that between the non-disinfected and disinfected SEV WWTP discharges at a 20 ppm.min PFA rate, a slight increase (less than 9%) in the average fluorescence intensity is noticed for all fluorescence peaks (α: 1.73 to 1.85 RU; β: 1.56 to 1.66 RU; α': 1.45 to 1.51 RU; δ: 1.26 to 1.32 RU; γ: 0.75 to 0.82 RU). The quality of SEV WWTP discharge fluorescent DOM (Figure 17b) is mainly composed of a 'humic substances-like' signal (α: 0.29 RU mgC $\cdot$ L^{-1}; β: 0.26 RU mgC $\cdot$ L^{-1}; α': 0.24 RU mgC $\cdot$ L^{-1}), followed by lower protein compound contributions (δ: 0.21 RU mgC $\cdot$ L^{-1}; γ: 0.12 RU mgC $\cdot$ L^{-1}). A small decrease (about 15%) in the average DOC-standardized fluorescence intensity is observed for all peaks between non-disinfected and disinfected samples treated at 20 ppm.min of PFA (Figure 17b). This variation is primarily due to an increase in the average measured DOC (DOC: 6.0 to 7.3 mgC $\cdot$ L^{-1},

+27%) caused by adding PFA. This phenomenon has also been demonstrated in Part 1.1 of this chapter.

In light of these results, any modification of the fluorescent DOM nature from SEV WWTP discharges was observed for PFA treatment rates between 0 and 20 ppm.min. It should also be noted that more PFA adds non-fluorescent DOC in solution, resulting in a decrease of total fluorescence intensities recorded per gram of carbon in solution. Moreover, it has been shown that this range of PFA treatment rates (0–20 ppm.min) exerts no influence on the total amount of fluorescent DOM observed in SEV WWTP discharge.

Analysis of SEV WWTP discharge EEM spectra according to extreme PFA processing rates (300 and 1000 ppm.min). Disinfection tests of SEV WWTP discharge with PFA concentrations of 300 and 1000 ppm.min were performed to determine the impact of high treatment rates on the DOM fluorescence signature. The fluorescence intensities obtained for two campaigns (Oct 30 and Nov 6, 2018) are shown in Figure 18.

The Oct 30 and Nov 6, 2018 campaigns display a similar fluorescent DOM nature (Figure 18a) mainly correlated with 'humic substances-like' fluorescent compounds (average fluorescence intensity of α: 2.02 RU), followed by the other DOM types (β: 1.69 RU; δ: 1.60 RU; α': 1.48 RU; γ: 1.19 RU). Let us note that unlike previous campaigns carried out (Figure 17), the measured non-disinfected SEV WWTP discharge contains a higher amount of 'protein-tryptophan-like' fluorescent compounds. Disinfection tests performed at 300 ppm.min PFA on SEV WWTP discharge do not significantly influence the average amount of fluorescent DOM observed, as opposed to the non-disinfected sample (relative standard deviation: less than 5% for all peaks). For a treatment rate of 1000 ppm. min PFA, decreases in the mean fluorescence intensity are mostly detected for 'protein-like' fluorescent components (γ: –33%, 1.19 to 0.80 RU; δ: –25%, 1.62 to 1.20 RU), in comparison with the other fluorescent components being monitored (α: –11%, 2.03 to 1.80 RU; β: –5%, 1.69 to 1.60 RU; α': +8%, 1.49 to 1.60 RU). One explanation for this phenomenon could be that the previous studies were carried out with PAA (Peracetic acid). A study by Domínguez Henao *et al.* (2018) demonstrated that proteins are the organic molecules with the greatest impact on the instantaneous demand for PAA due to their protein denaturation action. In addition, the study by Zhang *et al.* (2016) on the action of PAA on extracellular soluble compounds derived from sludge, from a membrane bioreactor fed by wastewater, also revealed a decrease in the fluorescence of 'protein-like' compounds. The oxidation action of PAA would thus result in the degradation of protein-like fluorescent compounds into smaller and potentially non-fluorescent molecules. The results found in our study for the 'protein-like' component in the case of significant PFA additions (i.e., above 1000 ppm.min) seem to be very consistent. It should be noted however that this range of treatment rates (PFA: 300–1000 ppm.min) has only been tested for exploratory purposes and will not be applied under actual operating conditions.

3.3 STUDY OF PFA INSTABILITY AFTER INJECTION

3.3.1 PFA degradation kinetics in WWTP discharge and surface water

3.3.1.1 Experimental description

For starters, long-term degradation kinetics were evaluated by monitoring over time the PFA concentration in a given PFA-doped matrix (SEV discharge water either unfiltered or filtered at 0.45 µm, or Seine River water filtered at 0.45 µm). The study matrix was temperature balanced for at least one night in a temperature-controlled chamber. Several temperatures were studied: 12, 20, and 25°C. One liter of water was then doped with PFA at a concentration between 1.5 and 2 ppm. Regular samples were then extracted every 10 min, and the PFA concentration was determined by means of ABTS, which is an alternative spectrophotometric method adapted from a peracetic acid determination (Pinkernell *et al.*, 1997). The ABTS method allows for high-frequency monitoring of PFA concentrations and thus yields a determination of low PFA concentrations. It is therefore well adapted to studying the decomposition of PFA and its monitoring after wastewater disinfection. Absorbance can be measured at two different wavelengths: 415 nm for colorless working samples, and 732 nm for colored working samples. The limit of quantification equals 0.05 ppm (at 415 nm, 1-cm light pass). The samples were stored in the temperature-controlled chamber until the absorbance measurement had been completed. Kinetic tracking continued for 90 mins.

Complementary experiments were then performed using raw SEV WWTP discharge samples in order to both identify the effect of fecal bacteria on PFA degradation kinetics and focus on the first 10 mins. A PFA dose of 1–1.2 ppm was applied to three distinct samples, and the PFA concentration was measured several times by the ABTS method with contact times ranging from 30 seconds to 10 mins. The experiments were performed on both raw and filtered (0.45 µm) samples.

3.3.1.2 Long-term degradation kinetics in WWTP discharge and surface water

Figure 19 shows the evolution in the concentration of PFA in SEV WWTP wastewater discharge (upstream) and in Seine River water (downstream). In both cases, decomposition kinetics were studied at three temperatures: 12, 20 and 25°C. To preserve sample homogeneity and prevent potential disturbance of absorbance readings due to the presence of TSS, the samples were filtered at 0.45 µm.

All kinetics have been correctly fitted by pseudo-first-order kinetics ($R^2 > 0.97$). The results allowed determining half-lives, in minutes, for both matrices at all three temperatures. These results are summarized in Table 3. For each matrix, the kinetics established at 20°C were performed in triplicate. The uncertainty was estimated at 10% and applied to the various temperatures studied.

For the two matrices analyzed, decomposition occurs faster as temperature increases. In SEV WWTP wastewater discharge, the half-lives varied from 33 mins (12°C) to 13 mins (25°C), whereas in Seine water the half-lives varied from 87 mins (12°C) to 29 mins (25°C). By way of comparison, Luukkonen *et al.* (2015) had obtained half-lives of 58 mins at 15°C in Finnish wastewater treated by physicochemical settling, biological treatment and tertiary filtration, under pseudo-first order kinetics.

It should be noted that the half-life of PFA in Seine water is less than 1 hour in summer (with a water temperature close to 20°C) and moreover that PFA is mostly decomposed in the WWTP discharge channel since the hydraulic residence time in this channel lies between 10 and 20 mins.

These results confirm the unstable nature of PFA compared to other peracids and particularly PAA since the half-life of PAA varies from 18 mins in primary effluent to 710 mins in tap water (Luukkonen & Pehkonen, 2017).

Let us also point out that the decomposition of PFA in wastewater showed quite significant variability at 20°C (half-life between 17 and 33 mins). This finding could probably be explained by variations in the quality of the water discharged (conductivity, TSS, DOC). Nevertheless, low concentrations of TSS do not seem to exert a major influence on the decomposition of PFA, as corroborated by the same half-lives of filtered and unfiltered samples (29 mins).

This decomposition reaction behavior is rather common and can be modeled by Arrenhius' Law ($\ln k = A + B \times (1/T)$, k = pseudo-first-order constant). The B factor in this model is directly related to the activation energy Ea of the studied reaction ($B = -Ea/R$, R = perfect gas constant), that is, the energy required for this chemical reaction to take place. In this case, the activation energy can be estimated at 57.8 kJ/mol for decomposition in Seine River water and 51.3 kJ/mol for decomposition in WWTP water. These values remain relatively low and are in agreement with the unstable nature of performic acid; they lie within the range of values found in the literature by various authors who have studied performic acid decomposition during the PFA synthesis process: 72.6 kJ/mol (Filippis *et al.*, 2009), and 52 kJ/mol (Santacesaria *et al.*, 2017).

3.3.1.3 Short-term degradation kinetics in WWTP discharge in the presence of fecal bacteria

Figure 20 displays the evolution in PFA concentration measured by the ABTS method in the three raw (non-autoclaved, non-filtered) samples of SEV WWTP discharge. The *E. coli* and intestinal enterococci concentrations in those samples were: 18,400 and 6,870 NPP/100 mL for Experiment 1, 40,600 and

11,200 NPP/100 mL for Experiment 2, and 51,700 and 13,800 NPP/100 mL for Experiment 3. The TSS contents were respectively 2, 10 and 12 mg/L, while the initial PFA concentration was 1–1.2 ppm.

PFA degradation kinetics are comparable in all three samples, with an initial instantaneous PFA consumption of 30–35% and then a slow concentration decrease. The PFA concentration is reduced by 40–50% after 10 mins in the presence of fecal bacteria. By means of extrapolation, it is possible to evaluate a half-life of 19–24 mins, which is highly comparable to the half-lives previously determined on autoclaved samples. This finding indicates that the presence of fecal bacteria does not significantly modify the PFA degradation kinetics in treated wastewater. Based on these results, it can be calculated that the initial PFA concentration is reduced by 58–69% after 20 mins, 73–87% after 30 mins and >90% after 50 mins. This calculation is most interesting since the actual PFA contact time observed at SEV during the industrial-scale trials (Part 2) was between 20 and 50 mins, with an average of 30 mins.

3.3.2 Analysis and fate of PFA degradation byproducts

The decomposition of performic acid was followed by liquid chromatography with both conductive detection (detection of ionized species such as acids) and UV detection (for detection of H_2O_2) of a solution containing about 100 ppm PFA, that is, a concentration 100 times higher than that actually used to generate favorable conditions for detecting potential byproducts. The only compounds detected were formic acid and hydrogen peroxide. These results are consistent with all published studies on the decomposition of performic acid, whereby the only stable products observed are formic acid, hydrogen peroxide and CO_2 (Filippis *et al.*, 2009; Leveneur *et al.*, 2014; Santacesaria *et al.*, 2017; Sun *et al.*, 2011).

During the experimental decomposition of PFA, the formation of hydroxyl radicals was also investigated by adding terephthalic acid (TA) to a PFA solution. TA specifically reacts with hydroxyl radicals to form a fluorescent compound, that is, hydroxyterephthalic acid (HTA) (Barreto *et al.*, 1994). Since no fluorescence signal of HTA has been detected, it can be concluded that the decomposition of PFA does not occur by means of homolytic rupture of the peroxide O-O bond and, consequently, no hydroxyl radicals are formed during the PFA decomposition process.

This result is in agreement with the low reactivity of PFA on DOM and organic micropollutants (see Section 1.2 of this chapter and Chapter 4) since it is well known that hydroxyl radicals react with DOM and numerous organic compounds (Wenk *et al.*, 2011).

Key points

- The increase in dissolved organic carbon in wastewater by PFA injection equals 0.78 ppm C/ppm PFA injected, which confirms the theory.
- 3D fluorescence highlights a negligible impact of PFA on fluorescent organic matter quality and quantity at a conventional PFA dose. A decrease in the 'protein-like' fluorescent organic matter is observed at a very high dose of PFA (1000 ppm.min).
- PFA degradation kinetics in SEV WWTP discharge and Seine River water follow a pseudo-first order with half-lives of 26 ± 9 min in wastewater and 53 min in Seine water at 20°C, after an instantaneous reduction of 30–35%.
- Temperature has a significant acceleration effect on PFA degradation, while TSS and fecal bacteria do not affect the degradation kinetics.
- The PFA degradation products are both formic acid and H_2O_2, with no other products being detected.

Chapter 4

The fate of micropollutants and byproduct formation during the disinfection of WWTP discharge by PFA

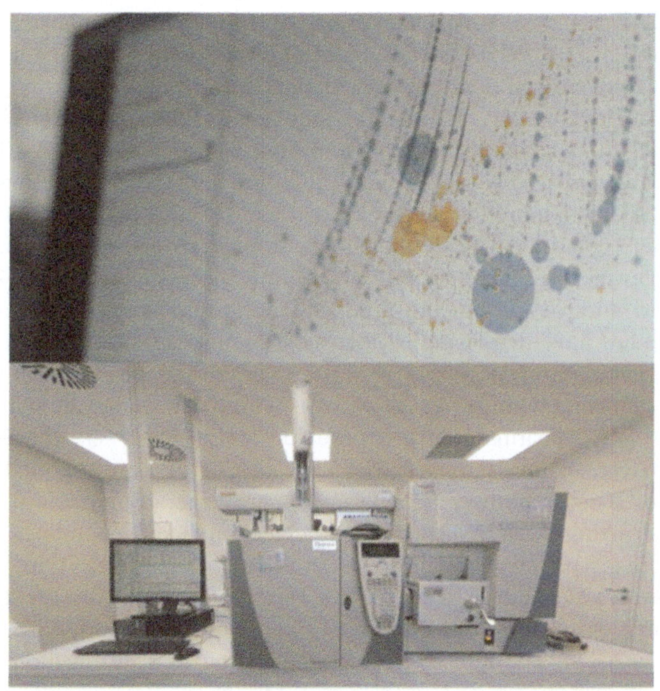

© Julien Le Roux, LEESU

doi: 10.2166/9781789062106_0039

4.1 INTRODUCTION

It is well documented that disinfectants commonly applied during water treatment processes (e.g., free chlorine, monochloramine, chlorine dioxide, ozone) can react with natural organic matter and inorganic ions (e.g., chloride, bromide, iodide) to generate disinfection byproducts (DBPs). The formation of DBPs is undesirable given that the majority of them have been characterized as carcinogenic and mutagenic (Wagner & Plewa, 2017). Since the discovery of chloroform formation from chlorinated organic matter in the 1970s, over 600 DBPs have been identified in disinfected water (Richardson *et al.*, 2007). However, a large proportion of the DBPs generated during disinfection processes remain unknown. It was reported that the identified DBPs only account for less than 50% of total organic halogens during chlorination (Krasner *et al.*, 2006). *In vitro* mammalian cell tests have demonstrated that nitrogenous DBPs (e.g., haloacetonitriles (HANs)) are more cytotoxic and genotoxic than trihalomethanes (THMs) and haloacetic acids (HAAs), which are commonly formed and regulated in drinking water (Plewa *et al.*, 2008). Generally speaking, brominated and iodinated DBPs are also more potent than their chlorinated analogues (Richardson *et al.*, 2007).

N-nitrosamines are a class of emerging nitrogenous DBPs of great concern due to their extremely high carcinogenicity. N-nitrodimethylamine (NDMA), which is the most frequently observed N-nitrosamine, presents a high risk to cause cancer with a probability of one in a million while exposed to a 0.7 ng/L solution (Mitch *et al.*, 2003). Its high polarity enables easy passage through reverse osmosis membranes during wastewater reclamation processes, thus necessitating the use of UV photolysis to remove it from the produced water supply (Mitch *et al.*, 2003). Previous studies have demonstrated that most N-nitrosamines originate from the dissolved organic nitrogen present in wastewater, and especially from secondary and tertiary amines (Schreiber & Mitch, 2007). Tertiary amines in anthropogenic compounds (e.g., pharmaceuticals, pesticides) (Le Roux *et al.*, 2011) and amine-based coagulation polymers (Park *et al.*, 2009) were reported as major precursors of N-nitrosamines. N-nitrosamines are generally present at low ng/L levels in drinking water, hence their detection requires robust analytical techniques (e.g., pre-concentration, tandem mass spectrometry) to reach low detection limits (Alexandrou *et al.*, 2018). Despite their typically low occurrence and concentrations, levels over 1000 ng/L were however reported in wastewater (Bond *et al.*, 2011). The reliance on chloramination as a disinfection process is known as a major source of N-nitrosamine formation. Break point chlorination (especially in the presence of nitrite ions), the ozonation of compounds containing dimethylamine groups and the sunlight treatment of nitrite-containing water are also known to generate N-nitrosamines (Shah & Mitch, 2012).

In recent years, peracids such as peracetic acid (PAA) and performic acid (PFA) have received increasing attention as alternatives to chlorine-based disinfectants (Domínguez Henao *et al.*, 2018). Previous studies have reported that high PAA concentrations ($>$30 mg/L) lead to the formation of brominated byproducts, including bromoform, dibromochloromethane and brominated phenols, in bromide ion-enriched solutions (Dell'Erba *et al.*, 2007; Shah *et al.*, 2015). PAA slowly reacts with chloride, bromide and iodide ions in water to generate secondary oxidants (i.e., hypochlorous (HOCl), hypobromous (HOBr) and hypoiodous acids (HOI)) in accordance with Equations (1)–(3) (k values were adapted from (Shah *et al.*, 2015)):

$$CH_3COOOH + Cl^- \rightarrow HOCl + CH_3COO^- \quad k = 1.47 \times 10^{-5}\,M^{-1}s^{-1} \quad (1)$$

$$CH_3COOOH + Br^- \rightarrow HOBr + CH_3COO^- \quad k = 0.24\,M^{-1}s^{-1} \quad (2)$$

$$CH_3COOOH + I^- \rightarrow HOI + CH_3COO^- \quad k = 4.2 \times 10^2 M^{-1}s^{-1} \quad (3)$$

These secondary oxidants can further react with organic matter to generate halogenated byproducts. HOBr is highly reactive with electron-rich moieties (e.g., activated aromatic rings, amines), with second-order rate constants up to three orders of magnitude higher than HOCl, and moreover forms brominated DBPs (Heeb *et al.*, 2014). However, under commonly applied water treatment conditions (PAA $<$10 mg/L), PAA produces less DBPs compared to chlorine or ozone (Kitis, 2004).

Furthermore, the residual H_2O_2 in peracid solutions can also play a role as an oxidant; it was reported that under acidic conditions, H_2O_2 also generates HOBr according to Equation (4) (Dell'Erba *et al.*, 2007):

$$H_2O_2 + Br^- + H^+ \rightarrow HOBr + H_2O \quad (4)$$

However, the H_2O_2 residual in water also acts as a sink for secondary oxidants; H_2O_2 reduces the free chlorine and bromine to chloride and bromide, respectively (Equations (5) and (6)) (Held *et al.*, 1978; Shah *et al.*, 2015):

$$H_2O_2 + HOCl \rightarrow Cl^- + {}^1O_2 + H_2O + H^+ \quad (5)$$

$$H_2O_2 + HOBr \rightarrow Br^- + {}^1O_2 + H_2O + H^+ \quad (6)$$

The presence of excess H_2O_2 in water may therefore reduce the formation of DBPs by consuming the secondary oxidants.

Similar to PAA, PFA is a powerful oxidant (Chhetri *et al.*, 2014) that can potentially react with organic matter and inorganic halides present in wastewater effluent. To the best of our knowledge, little is known about the potential formation of DBPs during the disinfection of wastewater effluent with PFA. A previous study reported that no DBPs (e.g., bromoform, detection limit: 0.5 µg/L) were found in wastewater disinfected with PFA (1 mg/L) and containing 0.4 mg/L bromide (Ragazzo *et al.*, 2013). Few studies have evaluated the formation of N-nitrosamines during PAA disinfection processes, but none regarding PFA processes. It was also reported that in the presence of 100 µg/L of amine precursors, 5 mg/L of PAA did not produce N-nitrosamines above the detection limit within 7 days of contact time, except for the formation of 8 ng/L of N-nitrosodipropylamine (NDPA) (West *et al.*, 2016).

In recent years, non-targeted analytical methods have generated interest in monitoring the evolution of organic compounds in the environment or during water treatment processes, thanks to the evolution of high-resolution mass spectrometry (HRMS). The goal of non-targeted screening is to detect as many ions as possible by means of HRMS in order to obtain an overview of the organic compounds present in a sample as well as to identify specific unknown molecules. Based on the accurate mass of each ion detected in a chromatogram, tentative molecular formulae are assigned and molecular structures can be derived, ultimately by searching in a library and injecting the corresponding analytical standard (if available) (Schymanski *et al.*, 2015). Chromatograms obtained by non-targeted analyses contain many compounds and thus generate large datasets that need to be examined using statistical tools. Such tools are commonly employed in metabolomics studies (Boccard *et al.*, 2010; Ramadan *et al.*, 2006) to: process a large number of variables, determine trends (e.g., increase of a signal over time), or identify molecules of interest in specific samples, for example, by searching for the compounds responsible for differentiating between groups of samples (Poulin & Pohnert, 2018; Schollée *et al.*, 2016). These methods have been used to assess the evolution of organic compounds during wastewater treatment processes (i.e., conventional treatment or advanced treatment, such as ozonation or adsorption onto activated carbon) (Bergé *et al.*, 2018; Merel *et al.*, 2017; Nürenberg *et al.*, 2019).

Suspect screening is another approach lying between targeted and non-targeted analyses. As in targeted analyses, a limited list of compounds is defined for suspect screening, based on various criteria such as the occurrence of compounds in previous studies or transformation mechanisms (Schymanski *et al.*, 2015). In contrast with the targeted analysis however, no reference standards are required for suspect screening, thus making it possible to search for compounds that are not commercially available (Krauss *et al.*, 2010). Consequently, confirmation and quantitation cannot take place during suspect screening, although the information related to the structure and molecular formula does prove to be helpful in proposing the presence of the compounds of interest in a sample analyzed by HRMS.

The primary purpose of this study has been to investigate the formation of DBPs during PFA-based disinfection of wastewater treatment plant (WWTP) discharge at both laboratory and industrial scales. Three classes of known DBPs were targeted: trihalomethanes (THMs), haloacetonitriles (HANs), and N-nitrosamines. Adsorbable Organic Halogens (AOX) were also analyzed in order to determine the formation of total halogenated organic compounds. The second objective of the study has been to characterize the evolution in organic micropollutants both before and after PFA disinfection by means of HRMS analysis. First, a non-targeted screening approach was applied to accomplish the global characterization of each sample, including total number of peaks and their total intensity, as well as the average retention time and average m/z of each peak. Second, a suspect screening was conducted based on a comprehensive list of 201 anthropogenic organic micropollutants (e.g., pesticides, pharmaceuticals, drugs), which had previously been detected in raw and/or treated wastewater (Bergé *et al.*, 2018; Mailler *et al.*, 2016, 2017).

4.2 TARGETED SCREENING OF DISINFECTION BYPRODUCTS

4.2.1 Experimental description

Wastewater effluent samples were collected between September 2018 and December 2018 from both the Seine Amont Valenton (SEV) WWTP and the Seine Centre (SEC) WWTP. Treated water (TW) at the SEV WWTP, corresponding to the PFA pilot inlet (see Part 2, Chapter 1) was collected on Sept 11, 2018, Sept 18, 2018 and Nov 6, 2018, and then used for laboratory scale disinfection experiments. Pretreated raw water (RW) and settled water (SW) were collected on Dec 11, 2018 from the SEC WWTP (see Part 1, Chapter 1) and used at the laboratory scale as well. Full-scale disinfection trials were performed on-site at the SEV WWTP; the samples were collected both before (SEV-TW) and after disinfection with PFA (i.e. disinfected water, SEV-DW) on Sept 26, 2018, Oct 10, 2018 and Oct 24, 2018. The experimental conditions applicable to each sample are given in Part 2, Chapter 1 for the full-scale disinfection trial samples. For the laboratory scale trials, 1, 2, 30 and 100 ppm of PFA were applied on TW samples, in addition to 1 ppm after injection of 2 mgN/L of nitrite and just 10 ppm of PFA for RW and SW from SEC.

All chemicals introduced were of analytical grade or higher and used as received without any further purification. A standard mix of chlorinated organics and DBPs, including four trihalomethanes (chloroform, dichlorobromomethane, dibromochloromethane and bromoform) plus three haloacetonitriles (dichloroacetonitrile, chlorobromoacetonitrile and dibromoacetonitrile), dichloropropanone and chloropicrin, was purchased from LGC Standards SARL (Molsheim, France) (100 ng/μL of each compound in acetone). Decafluorobiphenyl (99%) was supplied by Sigma-Aldrich Inc. (St. Louis, MO, USA). Ethyl acetate, Methyl tert-butyl ether (MTBE), sodium nitrate (99.99%) and nitric acid (70% v/v) were all supplied by Merck KGaA (Darmstadt, Germany). 2,4,6-trichlorophenol (98%) was purchased from Acros (Thermo-Fisher Scientific Inc., Waltham, WA, USA). The standard HCl solution (0.01 mol/L) used for the AOX-200 titration cell check was provided by Mitsubishi Chemical Analytech (Japan). The activated carbon for AOX measurements was purchased from Envirotech (Germany).

GC-MS grade dichloromethane (DCM) was purchased from Merck KGaA (Darmstadt, Germany). A mixed standard of N-nitrosamines, including N-nitrosodibutylamine (NDBA), N-nitrosodiethylamine (NDEA), N-nitrosodimethylamine (NDMA), N-nitrosodiphenylamine (NDPhA), N-nitrosodipropylamine (NDPA), N-nitrosomethylethylamine (NMEA), N-nitrosomorpholine (NMOR), N-nitrosopiperidine (NPIP), N-nitrosopyrrolidine (NPYR) (with 2000 µg/mL of each analyte in DCM) and Toluene-d8 (2000 µg/mL in DCM), were all purchased from Agilent Technologies Inc. (Santa Clara, CA, USA). The N-nitrosodimethylamine-d8 (NDMA-d8) and N-nitrosodipropylamine-d14 (NDPA-d14) (1 mg/mL in DCM) were both supplied by Cambridge Isotope Laboratories Inc. (Tewksbury, UK). Coconut charcoal SPE tubes were supplied by Sigma-Aldrich Inc. (Saint-Louis, MO, USA), and the ultrapure water was produced from Milli-Q IQ 7000 Merck KGaA (Darmstadt, Germany)).

4.2.2 Adsorbable organic halogens (AOX) analysis

AOX were analyzed according to the NF-EN-ISO-9562 standardized method. Fifty mg of activated carbon and 5 mL of $NaNO_3$ (0.2 mol/L) were added to 100 mL of wastewater sample, whose pH had been adjusted to pH <2 with nitric acid. After 1 h of constant stirring, the adsorbed activated carbon was filtered by ceramic frits with Activated Carbon Adsorption Unit SA-200 (Mitsubishi Chemical Analytech, Japan), followed by a nitrate-wash of activated carbon (25 mL, 0.01 mol/L of $NaNO_3$). AOX were then converted into hydrogen halides under high-emperature (950°) combustion of activated carbon with an Adsorbable Organic Halogen Analyzer AOX-200 (Mitsubishi Chemical Analytech, Japan). The hydrogen halide gas produced was then dehydrated and introduced into a titration cell. The limit of quantification in the titration cell was 0.7 µg of Cl. The AOX recovery rate of a standard solution (50 µg/L of 2,4,6-trichlorophenol) was found to equal 100.6%.

4.2.3 Halogenated DBP analysis

THMs and HANs were analyzed after liquid–liquid extraction according to EPA method 551. Samples (50–100 mL) were transferred into 120 mL amber glass bottles containing 10 g of NaCl. Fifty µL of decafluorobiphenyl (10 mg/L in acetone) were spiked as a surrogate standard and 3–6 mL of ethyl acetate were added as the organic solvent for most samples. The sampling bottle was vigorously and consistently shaken by hand for 4 min. The water and organic phases were allowed to separate for 2 min. About 2–3 mL of organic extract was transferred into a smaller vial and evaporated under N_2 gas. Lastly, 200 µL of concentrated extract was transferred into a GC-MS vial and stored at 4°C until analysis.

DBP quantification was performed in electronic ionization mode using a gas chromatograph (Trace GC Ultra, Thermo Fisher Scientific Inc., Waltham, WA, USA), coupled with a triple-quadruple mass spectrometer (TSQ Quantum, Thermo Fisher Scientific Inc., Waltham, MA, USA). One µL of extract was injected (inlet temperature: 200°C) in pulsed splitless mode. The compounds were separated on a Rxi$^®$ 5Sil MS (60 m × 250 µm × 0.25 µm) column (Restek, France). The oven program was held at 35°C for 9 min, ramping up to 125°C at 10°C/min, then increased to 220°C at 25°C/min and held there for 2 min. The total run time was 23.8 min. The DBP standard calibration curve was in the range of 50–500 µg/L in solvent, thus corresponding to a limit of quantification of 0.1 µg/L in water samples before extraction. The DBPs were quantified in single-ion monitoring (SIM) mode.

4.2.4 N-nitrosamine analysis

The extraction and detection method of N-nitrosamines was adapted from both US EPA method 521 and a gas chromatography method coupled with tandem mass spectrometry (GC-MS/MS) (Yoon *et al.*, 2012). One hundred µL of NDMA-d8 and NDPA-d14 (500 µg/L in DCM) were spiked as internal standards into 500 mL of a filtered (GF/F, 0.7 µm pore diameter) sample. The N-nitrosamines were extracted using coconut charcoal solid-phase extraction cartridges (6 g, Sigma-Aldrich Inc., Saint-Louis, MO, USA) using a Dionex™ AutoTrace (ThermoFisher Scientific Inc., Waltham, MA, USA) at a rate of 10 mL/min. The cartridges were conditioned with 5 mL of DCM, 5 mL of methanol and 10 mL of ultrapure water. After extraction, the cartridges were dehydrated for 30 min by purging the N_2. Twelve mL of DCM were used to elute the analytes. The eluate was concentrated by evaporation under a stream of N_2 until ~1 mL. Nine hundred µL of the concentrated extract were accurately transferred into GC-MS vials and stored at −20° until analysis. Forty-five µL of injection standard (Toluene-d8, 2 mg/L in DCM) were added into the extract before injection into GC-MS/MS.

The nine N-nitrosamines (i.e. N-nitrosodibutylamine (NDBA), N-nitrosodiethylamine (NDEA), N-nitrosodimethylamine (NDMA), N-nitroso-diphenylamine (NDPhA), N-nitrosodipropylamine (NDPA), N-nitroso-methylethylamine (NMEA), N-nitrosomorpholine (NMOR), N-nitrosopiperidine (NPIP) and N-nitrosopyrrolidine (NPYR), 2000 µg/mL each in DCM), plus the two internal standards (NDMA-d6 and NDPA-d14) and one injection standard (toluene-d8) were all analyzed in electron ionization mode using a gas chromatograph (Trace GC Ultra, Thermo Fisher Scientific, Waltham, MA, USA) coupled with a triple-quadruple mass spectrometer (TSQ Quantum, Thermo Fisher Scientific Inc., Waltham, MA, USA). One µL of extract was injected (inlet temperature: 250°C) in pulsed splitless mode. Chromatographic separation was completed in an Rxi$^®$ 5Sil MS (60 m × 250 µm × 0.25 µm) column (Restek, France). The oven program was held at 45°C for 3 min, ramping up to 130°C at 25°C/min, then increased to 230°C at 12°C/min and held there for 1 min. Total run time was 15.7 min. The calibration standards of N-nitrosamines were prepared in the range of 5–100 µg/L. Internal standards NDMA-d6 and NDPA-d14 were added into each standard as 50 µg/L. The toluene-d8 in calibration standards was 100 µg/L. The N-nitrosamines were quantified in selected reaction monitoring (SRM) mode using argon (6.0) as the collision gas. The detection parameters (collision energy for each compound) were optimized by injecting a 5 mg/L N-nitrosamine standard solution. Table 4 shows the selected analytical parameters for each N-nitrosamine (retention time, parent and product ions, and collision energies). The limits of quantification of N-nitrosamines were in the range of 0.1–6 ng/L.

4.2.5 Impact of PFA on AOX and halogenated DBP formation in WWTP discharge

Table 5 shows the concentrations of AOX and DBPs found in SEV or SEC wastewater before and after PFA disinfection processes, during both laboratory scale experiments and full-scale trials.

During all disinfection experiments (both at the laboratory scale and during full-scale on-site trials), >30 µg/L of AOX were already present in the wastewater effluent samples before any disinfection steps. At the laboratory scale, the disinfection of wastewater effluent samples using PFA concentrations of 2 and 30 ppm (C × t = 20 and 300 ppm.min, respectively) increased the AOX concentrations by 5–10 µg/L. The disinfection of raw and settled wastewater samples (December 2018) with 10 ppm PFA did not produce any significant amounts of AOX. A high concentration of 100 ppm of PFA (C × t = 1000 ppm. min, November 2018) significantly increased the formation of AOX from 45 to 339 µg/L, thereby indicating that halide ions (probably bromide ions) can be incorporated into organic matter upon reaction with high doses of PFA.

Table 6 presents the bromide ion concentration in SEV WWTP wastewater effluent both before and after full-scale disinfection by PFA (August–October 2018). Bromide ions were present in effluent within the range of 100–250 µg/L, and their concentration varied significantly over time. A decrease of about 10–40% in bromide concentration was observed after PFA treatment in the August–September (but not October) samples, which could be attributed to the reactivity of bromide ions with PFA (e.g., to generate HOBr-like PAA, see Equation (2)).

Among the regulated THMs, chloroform was not detected in any sample. Other halogenated DBPs were not detected in WWTP discharge samples either before or after disinfection using low doses of PFA (C × t = 10–30 ppm.min). At higher PFA doses (C × t = 74, 300 and 1000 ppm.min) and in disinfected raw and settled wastewater, dichlorobromomethane (DCBM), dibromochloromethane (DBCM), bromoform (TBM), bromochloroacetonitrile (BCAN) and dibromoacetonitrile (DBAN) were the most frequently detected DBPs. In most samples where DBPs were detected (except for the 1000 ppm.min PFA dose), DBP concentrations were determined to be less than 1 µg/L. The highest concentrations were observed for samples treated with PFA at 1000 ppm.min (e.g., DCBM = 2.4 µg/L, DBCM = 1.4 µg/L, BCAN = 2.0 µg/L, DBAN = 0.6 µg/L), in agreement with the high

AOX concentrations detected in this sample. Chlorinated DBPs were also formed (TCAN = 0.7 µg/L, DCP = 3.4 µg/L, TCNM = 1.8 µg/L), with DCAN reaching the highest concentration of all DBPs (8.25 µg/L). As shown in Equations (1) and (2), the rate constant of chloride with PAA is four orders of magnitude lower than that of bromide. The PFA rate constants with halide ions are not available. However, our results indicate that brominated DBP may be preferentially formed during PFA disinfection via the formation of HOBr as an intermediate, followed by the formation of HOCl and chlorinated DBPs at higher PFA doses, which would require subsequent confirmation. Furthermore, the formation rate constant of HOI during the PAA reaction with halides is roughly three and seven orders of magnitude higher than that of HOBr and HOCl, respectively (Equation (3)). Further investigation is therefore needed regarding the potential formation of HOI and iodinated DBPs, which may be favored during PFA disinfection. The high reactivity observed at 1000 ppm.min is also in accordance with results obtained using 3D fluorescence spectrometry (see Chapter 3), in showing a significant modification of the organic matter matrix. More specifically, a large decrease was observed for protein-like substances, which are known to be major precursors of nitrogenous DBPs such as HANs (Bond *et al.*, 2011).

An injection of 2 mg N/L of nitrite ions with a PFA dose of 10 ppm.min led to the increased formation of two DBPs (BCAN = 0.15 µg/L, DBAN = 0.10 µg/L), which were not detected in the absence of additional nitrite (November 2018), although the concentrations remained very low. A confirmation of the effect of nitrite ions on halogenated DBP formation would require additional experiments at various nitrite and PFA concentrations.

Low concentrations of DBCM, TBM and DBAN (<0.4 µg/L) were detected in both raw and settled water samples. A slight increase in these DBP concentrations was observed after disinfection of the raw water with 100 ppm.min PFA, but concentrations remained less than 0.5 µg/L.

During the on-site industrial disinfection of the WWTP effluent, no DBPs were detected at the lowest PFA doses (Sept 26, 2018 and Oct 10, 2018, C × t = 32.1 and 28.3 ppm.min, respectively), yet brominated DBPs were detected when a higher dose (C × t = 74.2 ppm.min) was employed, in agreement with results obtained from laboratory batch tests at higher doses (i.e. >100 ppm.min). In particular, an increase of ~0.4 µg/L in DBAN concentration was observed. As in batch experiments, AOX concentrations increased by 6–13 µg/L and were not significantly higher at the highest PFA dose (C × t = 74.2 ppm.min) than those at two other doses (C × t = 32.1 and 28.3 ppm.min).

Overall, the low doses of PFA generally employed to disinfect WWTP discharge (i.e., ~20–30 ppm.min) did not form any detectable amounts of halogenated DBPs, and moreover the concentrations observed at higher doses (>70 ppm.min) were always low compared to other oxidation processes and significantly lower than the regulatory thresholds defined for drinking water (total THMs <100 µg/L). In comparison, the chlorination of SEC WWTP discharge (2 ppm during 10 min in batch, March 2018) formed ~150 µg/L of AOX, while <10 µg/L of AOX were generated from PFA at a similar dose.

4.2.6 Impact of PFA on N-nitrosamine formation in WWTP discharge

The method performance from solid-phase extraction to GC-MS/MS analysis was monitored based on the recovery rate of NDMA-d6, which was spiked into each wastewater effluent sample with a known concentration (100 ng/L) using toluene-d8 as the injection internal standard. The average recovery was 76.5%, which is consistent with recoveries reported in US EPA method 521 (77.2%) for most N-nitrosamines on similar coconut charcoal cartridges.

Table 7 shows the concentrations of N-nitrosamines in WWTP discharge before and after treatment with PFA. Among the nine targeted N-nitrosamines, NMEA, NPYR and NDPhA were not detected in any of the samples. Less than 5 ng/L of NDEA, NDPA, NPIP and NDBA were found in most samples. Nineteen and 17 ng/L of NDEA and 27 and 30 ng/L of NDMA were present in the raw and settled water samples (December 2018), respectively. NDMA was the most abundant nitrosamine detected in the WWTP discharge (treated water) samples (19–33 ng/L, average: 28 ng/L), followed by NMOR (6–18 ng/L, average: 11 ng/L), at levels commonly reported in the literature. NDMA and NMOR were reported to be the most predominant N-nitrosamines in untreated and treated wastewater (Gerrity *et al.*, 2015; Krauss *et al.*, 2009). Industrial products, such as rubber, tires and dyes, are potential sources of NDMA and NMOR. NDMA can also be produced during cooking (Lee & Oh, 2016). Household products (e.g., laundry detergents, dishwashing soaps) contain morpholine (NMOR precursor) as impurities (Glover *et al.*, 2019).

Statistical analysis (t-tests) indicated the lack of any significant differences in N-nitrosamine concentrations after disinfection with PFA at any concentration rate, even at the highest dose employed (1000 ppm.min). However, the addition of 2 mg N/L of nitrite ions during the PFA disinfection (C × t = 10 ppm.min) of treated water significantly increased the NMOR level, from 14 to 162 ng/L (November 2018). Under acidic conditions, the nitrite ion is known to react with secondary amines to generate N-nitrosamines, but this nitrosation mechanism is very slow at neutral pH (Mirvish, 1975). However, studies have demonstrated that some carbonyl compounds (e.g., formaldehyde) can catalyze the nitrosation at higher pH (6.4–11), via the formation of an aldehyde adduct on secondary amines (Keefer & Roller, 1973). PFA also contains a carbonyl group in its structure and thus could catalyze a nitrosation reaction upon the addition of nitrite. As shown in Table 7, a slight increase in all other N-nitrosamines (except NPIP) also occurred after adding nitrite (November 2018), but to a much lesser extent than that of NMOR. This outcome was possibly correlated with the presence of high levels of NMOR precursors (e.g., morpholine) compared to other N-nitrosamine precursors. The NDMA concentration reached 30 ng/L after adding 2 mg N/L nitrite and 10 ppm.min PFA, as compared to the 21 ng/L observed in samples before and after disinfection with 10 ppm.min PFA.

The N-nitrosamine results obtained during on-site PFA disinfection of the WWTP discharge (Sept 26, 2018, Oct 10, 2018, Oct 24, 2018) were all in good agreement with results obtained at the laboratory scale. The concentrations of all N-nitrosamines were low (<19 ng/L), except for NDMA which reached 31–33 ng/L in both disinfected and non-disinfected effluents. As in the laboratory scale experiments, NMEA, NPYR and NDPhA were never detected. The concentration of certain N-nitrosamines (e.g., NDEA and NDBA) sometimes increased slightly after PFA disinfection, but this was not systematically observed across the three sampling campaigns, especially not during the campaign with the highest PFA dose (74.2 ppm.min). In considering the potential impact of nitrite ions on N-nitrosamine formation, as observed in laboratory scale experiments, it would have been worthwhile to check the nitrite concentrations in the three treated water samples in order to determine a correlation with the observed increases in NDEA and NDBA, but unfortunately the nitrite ion was not measured during these full-scale tests. NMOR did not show any significant increase during full-scale testing, which means that either its precursors were not present in the wastewater effluent or the nitrite concentration was insufficient to cause nitrosation.

Overall, no significant change was noticed for the NDMA concentration after PFA disinfection at any dose. This finding indicates that the WWTP treated water did not contain any significant concentrations of NDMA precursors and/or nitrite ions, and moreover that PFA does not produce significant amounts of NDMA, the most carcinogenic N-nitrosamine, from wastewater effluent.

4.3 NON-TARGETED INVESTIGATION OF MICROPOLLUTANTS DURING PFA DISINFECTION
4.3.1 Experimental description

All samples collected for DBP quantification were also used for non-targeted screening analyses. The samples were filtered using glass fiber filters with a 2.7 and 0.7-μm pore size (Whatman GF/D and GF/F, respectively) within 24 h of collection. One liter of each sample was acidified to pH 6 with sulfuric acid (98%); also, 40 μL of a mixed solution of internal standards (i.e. bisphenol A-d6, 4-n-octylphenol-d17, 4-octylphenol-diethoxylate and propylparaben-d4) were added to each sample. Solid phase extraction (SPE) is currently the most common technique to concentrate molecules, to ensure their improved detection by HRMS (Hogenboom *et al.*, 2009; Müller *et al.*, 2011; Singer *et al.*, 2016). Many studies have applied *universal* cartridges for SPE (e.g., OASIS HLB) in order to retain a large number of molecules (Ibáñez *et al.*, 2008). The combined use of various cartridges offers an alternative method to capture diverse substances with distinct physicochemical characteristics (Singer *et al.*, 2016). The samples in this study were extracted using an automatic SPE instrument (AutoTrace, Dionex Corporation, Sunnyvale, CA, USA) with cartridges containing a mix of four different phases (Oasis HLB – Water, ENV+ – Biotage, Strata X-AW – Phenomenex and Strata X-CW – Phenomenex). The cartridges were conditioned with 10 mL of methanol, followed by 10 mL of ultrapure water. After sample extraction, the cartridges were dried under a gentle stream of nitrogen for 30 min. The substances were then eluted with 6 mL of a mixture of methanol/ethyl acetate (50:50) and 1.43% ammonia (35%), followed by 3 mL of a mixture of methanol/ethyl acetate (50:50) and 1.7% formic acid. The extract was evaporated under a stream of nitrogen and reconstituted with 1 mL of ultrapure water and methanol (80:20). All extracts were then filtered with PTFE filters (0.2 μm) and transferred into vials for analysis. A sample of ultrapure water ('extraction blank') was extracted under the same conditions as the actual samples.

The samples were analyzed by means of ultra-high performance liquid chromatography (UPLC) coupled with an ion mobility spectrometer and a high-resolution time-of-flight mass spectrometer (IMS-QTOF) (Vion (Waters Corporation, Milford, MA, USA)). The analytes were separated by an ACQUITY UPLC-BEH C18 column (2.1 × 100 mm, 1.7 μm, Water). Ten μL of each sample were injected at a 0.45 mL/min flow rate. The mobile phase was constituted of: (A) ultrapure water +0.1% formic acid, and (B) acetonitrile +0.1% formic acid, while the gradient was 1 min isocratic with 98% A, a 25-min linear decrease to 2% A, 5 min isocratic with 2% A, and a 4-min equilibration time with 98% A. The UPLC column and samples were maintained at 40 and 10°C, respectively.

The ionization was operated by electrospray (ESI) in positive and negative mode with the following parameters: capillary voltage at 0.80 kV (2.50 kV in negative mode), source temperature at 120°C (100°C in negative mode), desolvation temperature at 500°C (250°C in negative mode), cone gas flow rate at 50 L/h, and desolvation gas flow rate at 1000 L/h (600 L/h in negative mode). A data-independent acquisition was performed in HDMS^{E} mode to obtain low (6 eV) and high (20–56 eV ramp) collision energy spectra to examine both the precursors and fragments. This acquisition was conducted between 50 and 1000 m/z with a 0.2 s scan time.

Each sample was injected in randomized triplicates to limit the intra-sequence variability caused by the instrument's analytical variability. Injection blanks consisting of mobile phase samples were injected along the sequence, along with a pool sample composed of equal volumes of each sample injected during the sequence. This pool sample was used to evaluate the analytical drift during the sequence. The UNIFI software (Water) was used for data acquisition and preprocessing, including 4D peak detection, isotopes and adduct clustering, in addition to the alignment of detected markers across samples. A list of markers with corresponding exact mass (m/z ratio), retention time, drift time (ion mobility) and intensity in each sample was generated by UNIFI. This list was also exported as a.csv file for further data analysis using the 'R' software (R Core Team, 2019) to produce global statistics or a *bulk characterization* (total number of markers, total intensity, average retention time, average m/z ratio) and visualizations for purposes of sample comparison. The markers that were unique or common to groups of samples were then isolated to produce Euler diagrams using the *limma* (Ritchie *et al.*, 2015) and *venneuler* (*v1.1-0*; Wilkinson, 2011) packages.

4.3.2 Suspect screening

Suspect screening was performed on the same dataset obtained by the non-targeted acquisition on UPLC-IMS-QTOF. A homemade library was created in the UNIFI software, comprising a list of 201 organic micropollutants with their exact mass (m/z ratio) and molecular structure. During the UNIFI processing step, the molecules of the library were searched in all chromatograms, targeted by their exact mass with a tolerance of 5 ppm. False positives were eliminated based on several criteria, namely: the shape of the detected peak, its retention time (based on the expected retention on the BEH-C18 column), and a fragments match (comparison of fragments obtained in high-energy spectra with fragment-generated *in silico* by UNIFI, as well as with fragments found in open access libraries).

4.3.3 Fate of micropollutants: bulk characterization by non-targeted screening

A total of 114,576 and 114,490 markers were obtained from the UNIFI detection algorithm among all 23 samples in positive and negative mode, respectively. The markers detected in all samples were processed to visualize differences between samples by plotting fingerprints based on their retention time, m/z ratio and intensity (Figure 21). This type of plot indicates a systematic decrease in marker intensity after PFA treatment for each type of sample. Modifications to the molecule type were also recorded: many high intensity markers observed at retention times longer than 15 min and with high molecular weights ($>400\ m/z$) disappeared, while other markers detected at shorter retention times (5–12 min) and lower molecular weights (200–400 m/z) exhibited an increase in intensity. This observation was even more pronounced at higher PFA doses (300 and 1000 ppm.min, data not shown), which indicates that PFA can decompose moderately hydrophobic compounds with high molecular weights and, to a lesser extent, produce low molecular weight molecules that are also more hydrophilic. This type of organic compound transformation is typical of other oxidation processes (e.g. chlorination, ozonation). Such information reveals that PFA, in addition to its disinfecting power, is able to both reduce the concentration of many organic compounds and modify to some extent the nature of molecules present in wastewater effluent. It is worth noting that the WWTP discharge sampled on Sept 18, 2018 (SEV-TW) exhibited a very different fingerprint (not shown), with very large peaks around 8 min, 400 m/z and nearly no difference (i.e. limited decrease in intensity) after a PFA treatment at 20 ppm.min. This outcome was attributed to a degraded water quality due to a bypass of the tertiary phosphorus removal step that occurred in the SEV WWTP during a rain event on Sept 18, 2018 (see Part 2, Chapter 1). Total phosphorus in this sample was 2.5 mgP/L vs. 0.8–1.0 mgP/L under nominal conditions (e.g., SEV-TW from Sept 11, 2018 and Nov 6, 2018). The absence of the coagulation step normally used for phosphorus removal may also degrade water quality through the presence of colloidal matter capable of adversely affecting PFA effectiveness, leading to a degradation of organic compounds. The presence of colloids can also affect the sample preparation step (i.e., SPE), which was reflected in the strong variability observed between replicate fingerprints of this lone sample (i.e., higher variations between marker intensities of the three replicate injections).

A bulk characterization of each sample was performed in order to obtain average information from an entire fingerprint: average retention time, average m/z ratio of the detected markers, plus total number and total intensity (sum of intensities from each individual marker) of all detected markers. This characterization step was first completed with no prior filter (i.e. no threshold on intensities) for each treated water sample before any PFA disinfection (all six SEV-TW samples) and after PFA disinfection at either the full scale (three SEV-DW samples) or the laboratory scale with 20 ppm.min PFA (SEV-TW + 2 ppm PFA). The results obtained for the six samples before and after PFA disinfection were averaged and are summarized in Table 8. The average m/z ratio decreased after PFA treatment, thus supporting the previous observation that organic compounds had a lower molecular weight after oxidation. Retention times and the total number of markers on the whole remained stable, while the total intensity slightly decreased (from 4.8×10^8 to 4.1×10^8), which confirmed the overall reduction in peak intensities visually observed on the fingerprints (Figure 21). These parameters were also calculated from markers detected only before *or* after PFA disinfection (i.e., by removing common markers present both before *and* after PFA treatment). This calculation also clearly confirmed that molecules disappearing after PFA treatment had a higher molecular weight than organic compounds formed after PFA addition, yet the overall slight increase in average retention times did not support the expected formation of more hydrophilic degradation products. The number of molecules appearing after treatment (2,119) was higher than that of the disappeared compounds (1799), but their overall intensity was lower (4.1×10^7 vs. 4.5×10^7).

An intensity value threshold ($>10,000$) was applied to filter out the smallest peaks that were closer to noise peaks (the intensity values of individual markers ranged between 50 and 41×10^6, and peaks with intensity $<10,000$ were mainly present in blank samples). Approximately 60% of all markers were still present after applying this threshold. The total number of markers slightly decreased (from 7369 to 6222, see Table 8) after PFA treatment, as did their total intensity. When considering unique markers, the number of markers formed after PFA treatment (1303) was lower than the number that were totally decomposed (1792), while their overall intensity was slightly lower.

Euler diagrams (Figure 22) allow visualizing the number of markers exclusively present before or after PFA disinfection (i.e., unique markers in Table 8), as well as the number of markers present both before and after treatment (with no indication of their intensity). A clear decrease in the number of markers is observed after treatment (20 ppm.min of PFA) in both positive and negative ionization modes. A small proportion of new peaks was generated during the treatment and can be labeled as decomposition products (i.e. ~18 and ~10% of all peaks present after PFA treatment in positive and negative mode, respectively). A substantial proportion of markers initially present in the WWTP discharge were still present after PFA treatment (i.e. ~34 and ~52% in positive and negative mode, respectively). Similar Euler diagrams were obtained from other samples (data not shown), thus confirming the qualitative observations made from marker fingerprints (Figure 21) that PFA treatment reduces the number of organic compounds detected in the samples.

The overall slight decrease in the average number of markers across all samples after PFA disinfection (Table 8) is detailed for each sample in Figure 23. Some samples (e.g., on Sept 11, 2018 or Nov 6, 2018) exhibited a clear decrease in the number of markers, while all other samples showed no decrease or even a slight increase (especially on Sept 18, 2018, Oct 10, 2018 and Oct 24, 2018). These differences could be attributed to the occurrence of bypasses in the WWTP during the rain events that occurred on Sept 18, 2018, Sept 26, 2018 and Oct 10, 2018, and thus to variations in the WWTP discharge matrix. However, an increase in the number of markers was also observed on Oct 24, 2018 on a date when no bypass occurred, so other unknown factors may have played a role. These results show that the reactivity of PFA with organic molecules can be quite variable depending on underlying conditions (e.g., quality of the wastewater matrix, presence of inorganic ions, temperature). In most cases, the impact of PFA on the number and intensity of markers was very limited. On Nov 6, 2018, the number of markers decreased with an increase in PFA concentration to 1 and 2 ppm, but then increased again at higher PFA doses (30 and 100 ppm of PFA). This finding can be attributed to the formation of additional degradation products, which were indeed observed in the lower range of molecular weight and retention time on fingerprints; it is also in accordance with the higher occurrence of AOX and halogenated DBPs analyzed in these samples. Nonetheless, the number of markers did not exceed that of the initial WWTP discharge. The presence of nitrite ions seemed to decrease the number of markers detected after treatment with 1 ppm PFA, but this assertion would require further confirmation. Lastly, the disinfection of raw (SEC-RW) and settled (SEC-SW) wastewater (Dec 11, 2018) did not reveal any significant difference in the number of detected markers.

Similar results were obtained by calculating the total intensity of markers in each sample (data not shown), thus demonstrating that the overall concentration of molecules slightly decreased after PFA disinfection and moreover that molecules produced during PFA treatment were not of high intensity. The overall limited decrease in number of markers, as well as the limited formation of organic compounds at low retention times (i.e., more hydrophilic molecules) as described earlier, is in accordance with the low reactivity of PFA in forming DBPs; it is also in agreement with the low impact observed on the organic matter matrix by means of fluorescence spectroscopy measurements (Chapter 3).

4.3.4 Identification and fate of micropollutants characterized by suspect screening

A suspect screening approach was employed on HRMS datasets in order to study the impact of PFA disinfection on specific organic compounds detected in the samples. This suspect screening was performed on a list of 201 anthropogenic organic micropollutants (e.g., pesticides, pharmaceuticals, drugs) that had been previously detected in raw wastewater and/or treated water before discharge (Bergé *et al.*, 2018; Mailler *et al.*, 2016, 2017).

Many of the listed molecules were detected in at least one sample (as revealed by the number of molecules detected in the pool sample). No significant differences were observed between the number of molecules detected before and after PFA disinfection, even at the highest doses (300 and 1000 ppm.min). The intensity of the selected molecules was isolated and corrected by internal standard signals (propylparaben-d4 intensity in each sample) in order to compensate for any deviation between samples. Figure 24 describes the evolution in the pharmaceuticals acetaminophen and sulfamethoxazole, as well as that of 11-nor-9-carboxy-tetrahydrocannabinol, the main metabolite of the psychoactive substance tetrahydrocannabinol (THC) in the samples. These molecules were detected at a high occurrence and relatively high concentrations in raw (e.g., acetaminophen $>100\,\mu g/L$, sulfamethoxazole $>500\,ng/L$) and treated wastewater (Guillossou *et al.*, 2019). These three molecules exhibited a decrease in intensity in most disinfected samples. This result illustrates the moderate degradation of organic micropollutants by PFA and furthermore explains that most molecules were still detected even after PFA treatment.

The individual intensities of each organic compound were averaged by category of molecule, then normalized and reported at increasing PFA doses (0, 10 and 20 ppm.min) from the WWTP discharge sample of Nov 6, 2018 (Figure 25). Once again, most families of molecules exhibited a slight decrease in average intensity at 10 ppm.min and remained stable as the PFA dose was increased to 20 ppm. min. Some categories (e.g. radiocontrast agents, hormones, antibiotics) showed higher removals, with an average intensity decreasing by more than 50%. Alkylphenols displayed a slight decrease at 10 ppm.min of PFA, but a major increase when 20 ppm.min of PFA were used. This finding could be attributed to the degradation of long-chain alkylphenols and their ethoxylates present in treated wastewater (originating from agricultural, industrial and household products, Bergé *et al.*, 2012), which could react with PFA to produce the simple alkylphenols (e.g., 4-nonylphenol and 4-tert-octylphenol) listed and used in this suspect screening approach. All these results should be carefully considered as qualitative information and would require confirmation by targeted analyses of the organic micropollutants (i.e., a precise quantification by using analytical standards and calibration curves).

Key points

- Disinfection of the WWTP discharge with moderate doses of PFA (10–30 ppm.min) did not generate any detectable amounts of known halogenated DBPs (e.g., THMs, HANs). AOX concentrations generally increased by 5–10 µg/L from the initial 30–40 µg/L concentrations in WWTP discharge. Higher PFA doses (>70 ppm.min) generated low concentrations of halogenated DBPs (DCBM, DBCM, TBM, BCAN and DBAN), which were well below the regulatory threshold for drinking water (e.g. European Union Standards: total THMs = 100 µg/L).
- Low concentrations of N-nitrosamines were detected in the WWTP discharge, with NDMA being the most abundant (up to 33 ng/L). On the whole, no significant change was noticed in the concentration of any N-nitrosamine after PFA disinfection at any dose, which indicates that the WWTP treated water did not contain significant concentrations of N-nitrosamine precursors and/or that PFA is not capable of producing N-nitrosamines. A large increase in NMOR concentration was however detected in the presence of nitrite ions (2 mg N/L) at a low PFA dose (10 ppm.min). This finding should be confirmed by mechanistic studies using NMOR precursors (e.g., morpholine).
- Non-targeted screening by HRMS showed a low reactivity of PFA with organic compounds. Some decomposition of organic molecules was observed even at low PFA doses but was not responsible for producing a substantial number of new molecules (i.e., degradation products), hence in agreement with the low formation of DBPs. Higher PFA doses did increase the number of molecules detected by HRMS, as compared to the low doses by generating degradation products, yet this production never exceeded the number and total intensity of signals observed in the initial WWTP discharge.
- Selected organic micropollutants were just slightly decomposed by commonly employed doses of PFA (10–30 ppm.min). Most categories of

micropollutants exhibited a low reactivity with PFA. Some categories (e.g., radiocontrast agents, alkylphenols) displayed higher reactivity.

- Forthcoming work on HRMS datasets will be dedicated to identifying unknown molecules specific to the wastewater disinfected by PFA, through the use of statistical tools (e.g., principal component analysis, orthogonal partial least squares discriminant analysis) to potentially identify degradation products of interest.
- The overall low reactivity of low doses of PFA towards both the degradation of organic compounds and the formation of DBPs was in agreement with the low reactivity of PFA with the organic matter matrix, as observed by fluorescence spectrometry (Chapter 3).

Section 2

Effectiveness of the Chemical Disinfection Process at Full Scale (Seine Valenton WWTP)

Chapter 1

Technical description of the industrial trials conducted at Seine Valenton WWTP

© SIAAP

1.1 INTRODUCTION

In order to verify the industrial-scale effectiveness of PFA disinfection and its impact on Seine River water, industrial-scale trials were performed in 2018

doi: 10.2166/9781789062106_0059

within the Paris Metropolitan Area. The identified recreation spots were located in the center of Paris, near the Ile de la Cité and Tour Eiffel. The SIAAP wastewater treatment plant (WWTP) discharges located upstream impacting these sites are Seine Valenton (SEV) WWTP and Marne Aval (MAV) WWTP. SEV treats far more wastewater than MAV (by a factor of 10), and MAV already had a UV disinfection unit, so the PFA industrial-scale trials were conducted at SEV.

Few industrial-scale trials were found in the literature, especially on WWTP treated water discharged into a river. Among them, Venice (Italy) in 2005–2011 (Ragazzo *et al.*, 2013), Biarritz (France) in 2015–2017 (Pigot *et al.*, 2019) and Berlin in 2012 (internal report, unpublished) can be cited for purposes of comparison, but they concern wastewater discharged into the sea, a lagoon or lakes. The Venice, Biarritz and Berlin cases will be technically described in Part 4 of this book. In addition, PFA disinfection units have been industrially implemented to disinfect combined sewer overflows in Denmark (Chhetri *et al.*, 2015).

Considering that SEV WWTP treats around 550,000 m^3/day under normal conditions, as explained in this chapter, such industrial-scale PFA disinfection trials are, to the best of the authors' knowledge, the most extensive performed thus far in the world. Moreover, the Paris Metropolitan Area case is very interesting since during summer (recreational period), the SEV WWTP discharge accounts for around 6 m^3/s, while the Seine River in summer normally flows at around 80–100 m^3/s. The anthropic pressure exerted by Parisians on the Seine River is thus extremely strong during the summer season, which makes achieving adequate bathing quality a real technical challenge.

This chapter will present in detail a technical description of the industrial trials performed at SEV WWTP in 2018. The process layout of the plant will be described first, followed by the tested Kemira Desinfix technology. Lastly, the trial design will be explained, including the PFA disinfection conditions applied, the sampling methodology and the analytical parameters monitored.

1.2 PRESENTATION OF THE SEINE VALENTON WWTP

The Seine Valenton (SEV) WWTP is located 15 km southeast of Paris and has a treatment capacity of 600,000 m^3/day (2,600,000 population equivalent) under normal conditions and between 600,000 and 1,500,000 m^3/day under degraded conditions, including wet weather. This WWTP is capable of receiving a maximum peak flow rate of 21 m^3/s; it treats wastewater originating mainly from the eastern part of Paris before discharging the treated water into the Seine River, upstream of the city of Paris. The WWTP layout is composed of two parallel treatment lines, named Valenton 1 and Valenton 2. Figure 26 summarizes the SEV WWTP treatment layout.

Raw wastewater is first pretreated by screening and grit and oil removal. Next, the pretreated wastewater is split between Valenton 1 and Valenton 2. Under normal operating conditions (SEV flow rate below 550,000 m^3/day), Valenton 1 treats the pretreated wastewater by primary settling before a biological treatment using low-charge activated sludge with a succession of anoxic and anaerobic tanks, aimed at treating carbon and nitrogen, along with a tertiary physicochemical treatment adding $FeCl_3$, anionic polymer and micro-sand. Under degraded conditions, the tertiary treatment is partially bypassed; moreover, under normal operating conditions, Valenton 2 treats the pretreated wastewater by primary settling before a biological treatment by means of low-charge activated sludge with a succession of endogenous, anoxic, anaerobic and aerobic tanks for treating carbon and nitrogen. Under degraded conditions, primary settling is partially bypassed. In wet weather, when the raw wastewater flow exceeds 11.4 m^3/s, a fraction of the pretreated wastewater is treated only by the tertiary physicochemical lamellar settling units. Above 17.6 m^3/s, all tertiary units are treating pretreated wastewater, and above 21 m^3/s or 550,000 m^3/d, the tertiary units are bypassed and pretreated water is discharged directly into the Seine River.

1.3 PRESENTATION OF PFA DISINFECTION BY APPLICATION OF THE KEMIRA KEMCONNECT DEX TECHNOLOGY

The disinfection technology tested in the SEV WWTP is the Kemira Desinfix process, based on *in situ* performic acid (PFA, DEX-135) production and injection. PFA is obtained by mixing hydrogen peroxide (DEX-550) and catalyzed formic acid (DEX-A375); it is highly unstable, thus making its production necessary on-site just before injection. Consequently, this technology basically consists of a mixing unit and two storage tanks (Figure 27). The PFA output contains approximately 13.5% PFA by mass, 20% H_2O_2 and 30.9% residual formic acid.

1.4 DESIGN OF THE INDUSTRIAL-SCALE TRIALS
1.4.1 Applicable PFA disinfection conditions

The trials were performed between late August and late October 2018, with a total of 10 sampling weeks. Five weeks were dedicated to evaluating PFA disinfection effectiveness and the other five weeks were spent monitoring as a control without PFA injection. The PFA dose injected during the trials is given in Table 9.

1.4.2 Sampling methodology

Throughout the 10-week period, the same sampling methodology was applied, including treated, disinfected and Seine River samples. Figure 28 provides the exact location of the sampling points as red crosses plus a photo of each point.

The samples were collected upstream (1) and downstream (2) of the PFA injection in order to evaluate the removal effectiveness of the treatment.

The downstream samples were collected at the end of the WWTP discharge channel (3.65 km from the PFA injection point), corresponding to a hydraulic retention time (and PFA contact time) before discharge to the Seine River of about 10–30 min, depending on hydraulic conditions. Point-specific samples were collected three times a day and 3 days a week (Tuesdays, Wednesdays and Thursdays), with a telescopic rod for the upstream point and a pump for the downstream point; a gap of 30 min was introduced between upstream and downstream samples.

Other samples were collected from the Seine River both upstream and downstream of the WWTP discharge point to evaluate the impact of WWTP discharge on the river with and without PFA injection. These samples were collected at two spots of the MeSeine Platform, which is responsible for Seine River quality monitoring in the SIAAP Authority: Choisy-le-Roi (PK 622.440) (3) and Port à l'Anglais (PK 626.152) (4). The Choisy-le-Roi sampling point is located approximately 1500 m upstream of the WWTP discharge point, on the first bridge accessible to pedestrians. The Port à l'Anglais sampling point is located in Alfortville about 2500 m downstream of the discharge point, also on the first bridge accessible to pedestrians. At both sampling points, one-time samples were collected on the right bank of the Seine River, the left bank and the middle using a bucket and a rope thrown from the bridge; the samples were then mixed together with equal volumes. Sampling was performed once a week (Wednesdays). It is important to keep in mind that depending on weather conditions, rainfall sewer overflows can occur on occasion between these two Seine River sampling points, significantly lowering water quality in the process.

1.5 ANALYTICAL PARAMETERS MONITORED

Several parameters were analyzed within the collected samples; they are given in Table 10 along with the frequency and type of water in which their determinations took place.

Regarding pathogens, *E. coli* and intestinal enterococci were analyzed in all samples, while spores of anaerobic sulfite reduction bacteria (SSR) and F-specific RNA bacteriophages were analyzed solely in the WWTP discharge and disinfected water once a week (Tuesdays). Conventional wastewater quality parameters were also assessed in all samples, including total suspended solids (TSS), carbon (dissolved organic carbon, chemical and biochemical oxygen demands), nitrogen (Kjeldahl nitrogen, ammonium, nitrite and nitrate) and phosphorus (total phosphorus and orthophosphates) parameters, as well as pH, conductivity and turbidity. Specific analyses were performed in SEV discharge and disinfected water once a week (Tuesdays) to determine the halogenated organic hydrocarbons (AOX), bromide and bromate. Similarly, color, chloride and sulfate were measured only in the Seine River samples.

All analyses were conducted by the central SIAAP Laboratory or the CARSO Laboratory, in accordance with the reference methods. The analytical methods, limits of quantification and estimated analytical uncertainties are listed in Table 11.

Lastly, several operations-related datasets from SAV WWTP or the Seine River were collected to process the results. For SEV WWTP, this dataset included the flow of SEV WWTP discharge at the time of sampling, the water level and geometry of the discharge channel, and information regarding the internal process bypass within the WWTP capable of impacting discharge quality. For the Seine River, this dataset included the daily flow of this section of river collected from the online HYDRO databank (http://hydro.eaufrance.fr), precipitation collected from the SAV WWTP monthly operations report, and Seine River temperature collected from the MeSeine databank at the Alfortville monitoring point.

In situ monitoring of fecal bacteria at the Seine Valenton WWTP using ALERT rapid microbiology instrumentation (Fluidion®)

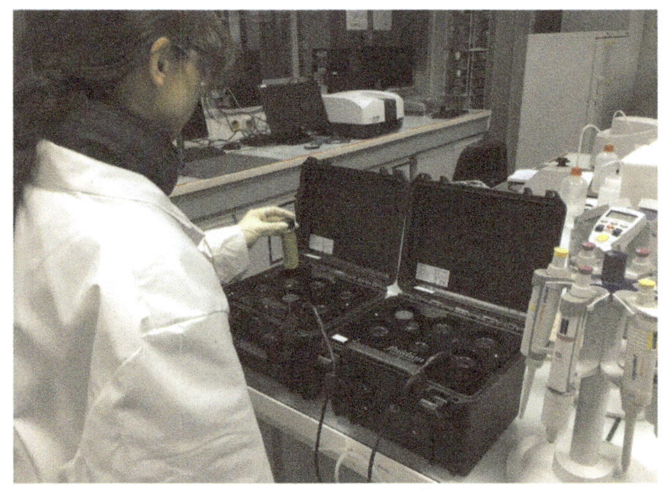

© SIAAP

2.1 INTRODUCTION

Identifying the presence of fecal indicator bacteria and measuring their concentration is a critical aspect of water quality monitoring, exerting a direct impact on safety aspects related to water usage: a safe drinking water supply,

doi: 10.2166/9781789062106_0064

recreational and competitive uses (swimming, boating, aquatic sports), agricultural use (irrigation), as well as aquaculture operations. Exposure to fecal pathogens through contact with contaminated water constitutes a major health risk, causing a wide variety of illnesses and infections with potentially fatal consequences, and moreover is recognized by the World Health Organization as such a health risk (WHO, 2003). Microbiological pollution with human and animal waste pathogens can be caused, in urban areas, by bacteria from wastewater plant effluent. This effect can be amplified during heavy rain episodes by combined sewer overflow (CSO) phenomena and, in some cases, by illegal discharges, boat sewage and faulty connections to the sewage infrastructure. In rural settings, agricultural runoff (e.g., from livestock operations) as well as the natural presence of birds and other warm-blooded animals are often responsible for microbiological pollution.

Monitoring the microbiological quality of drinking water and surface water in sensitive areas (i.e., where direct human contact or food chain contamination can create a health risk) is mandatory throughout the industrialized world and typically performed by measuring concentrations of viable and cultivable fecal indicator bacteria (FIB). These bacteria are generally not considered to be illness vectors by themselves, but rather provide an accurate tracer or proxy (Prüss, 1998). Depending on the country, the choice of specific FIB used for monitoring may vary: *E. coli* is preferred for general public health protection in fresh water (Edberg *et al.*, 2000); intestinal enterococci are more prevalently monitored in seawater environments, where they often complement or even replace *E. coli* measurements; and Thermotolerant Coliforms and Generic Total Coliforms are used in certain parts of the world, or for specific water matrices.

Typical standard regulatory FIB monitoring methods tend to follow one of two approaches. The more common approach is the most probable number (MPN) method, which employs either micro-assays (e.g., 96-well microplates (ISO 9308-3 1998), as the method used in many European countries) or larger-scale assays (e.g., Quanti-Tray 2000 (Quanti-Tray is a trademark or registered trademark of IDEXX Laboratories, Inc. or its affiliates in the United States and/or other countries) (ISO 9308-2 2012), now widespread in the United States, or multiple-tube fermentation (EPA 9131), now mostly obsolete). The second approach consists of membrane filtration followed by plating on a chromogenic Agar medium; it is generally reserved for low-concentration samples, such as in drinking water studies (ISO 9308-1 2014) These standard regulatory methods are quite often slow, with a typical time-to-response (TTR) being between 24 and 72 hours when taking into account all sampling, transport, storage and quantification protocol steps. Many situations require significantly faster early-warning systems in order to impose access closures or adapt operations accordingly. This need is particularly important wherever high risks of human contact with contaminated water are involved, such as when discharging upstream of a sensitive area (e.g., drinking water intake, recreational activities, irrigation of fresh produce, various forms of aquaculture). A strong need exists in the water and environmental engineering industry for rapid measurement devices capable of performing a reliable *in situ* quantification of the viable indicator bacterial load.

Previously reported online quantification techniques have generally focused on *E. coli*, while some have been applied to surface water quality monitoring (Lopez-Roldan *et al.*, 2013; Noble & Weisberg, 2005). Attempted enumeration methods range from simple light-scattering measurements to direct color and fluorescence measurements (Baker *et al.*, 2015), including complex molecular techniques. Certain methods, such as Reverse-Transcription Quantitative PCR (Bergeron *et al.*, 2011) and direct measurement of enzymatic activity (Baudart *et al.*, 2009; Briciu-Burghina *et al.*, 2015; Burnet *et al.*, 2019; Heery *et al.*, 2016; Wildeboer *et al.*, 2010), can provide initial results very quickly, in as little as 4 hours; however, such methods are prone to counting both cultivable and non-cultivable bacteria and are easily interfered by other types of microorganisms or by free enzymes present in the sample. This observation is particularly true of enzymatic techniques without a selective growth step; moreover, it becomes highly relevant whenever such methods are performed in a wastewater environment, where disinfection techniques may deactivate, but not kill, bacteria and where high background levels of free enzymes exist. For all these reasons listed, such methods do not generally provide a very robust correlation with regulatory measurements focusing specifically on viable, cultivable cells (Burnet *et al.*, 2019). Fluorescent *in situ* hybridization has been applied to bacterial detection as well (Baudart & Lebaron, 2010), although it remains limited to laboratory usage due to significant protocol complexity. Defined Substrate Technology (DST), combining a selective growth medium for the bacteria of interest with enzyme substrates linked to specific chromogens and/or fluorogens produced by bacterial metabolism, stands out as a reliable detection technique. Several approved quantification methods combining DST assays with MPN techniques also exist and have been standardized (ISO 9308-2, ISO 9308-3).

The Fluidion® ALERT line of instrumentation for monitoring microbiological contamination is a novel technology that utilizes a modified real-time DST method for bacterial enumeration. Fluidion® ALERT technology allows for a fully-automated *in situ* quantification of viable and cultivable generic *E. coli* and total coliforms or, alternatively, of the intestinal enterococci concentration or fecal coliform concentrations in both fresh water and seawater environments. The TTR ranges between 2 and 12 hours (shorter response times correspond to higher concentrations), and the limit of detection corresponds to one target bacterium in the 25-mL sample volume (i.e., a limit of detection of four bacteria/100 mL in fresh water, which if applied for specific protocols would need to be multiplied by the pre-dilution factor). An upper measurement range of 5×10^5 bacteria/100 mL is obtained with no need for serial sample dilution, thus covering in a single measurement over five orders of magnitude in concentration. The detailed calibration and metrological validation results for fresh surface water *E. coli* enumeration have been published elsewhere (Angelescu *et al.*, 2018a). The fact that no sample transport, conditioning or preparation is necessary yields certain logistical advantages and eliminates the risk of sample degradation and human error present in current techniques, while providing rapid and reliable information on water quality to enable effective decision-making. ALERT technology has been employed in numerous applications worldwide, ranging from seawater monitoring to the *in situ* environmental monitoring of highly-polluted streams (Angelescu *et al.*, 2018b), including source-identification investigations performed by regulatory agencies (Cronin *et al.*, 2018; Loewenthal *et al.*, 2018) and non-profit organizations (Angelescu & Saison, 2020) and implementation on various platforms, notably in conjunction with remotely-controlled aquatic drones (Angelescu & Hausot, 2019).

Since 2017, Fluidion® ALERT technology has been deployed at multiple sites along the Seine River in Paris to perform high-frequency rapid *E. coli* quantification in river water, thus complementing long-term monitoring efforts pursued by multiple entities involved in the short water cycle (City of Paris, SIAAP (SIAAP: *Syndicat Interdépartemental d'Assainissement de l'Agglomération Parisienne* is the Greater Paris Sanitation Authority) and neighboring departments) aimed at establishing environmental baselines and quantifying the impact of various mitigation actions adopted as early as 2000. Initial mitigation efforts, aimed at limiting phosphorus and nitrogen influx and reducing sewer overflow during rainstorm events, have resulted in continuous improvements to the microbiological quality of the Seine River over the past few decades (Rocher & Azimi, 2016).

Recent years have witnessed strong renewed interest in natural bathing sites, not only along the coast but also in urban environments, by both public authorities, through increasingly considering urban rivers as recreational resources, and the public, through various grass-roots actions (Ziegler, 2019). Competitive sports are, in turn, becoming increasingly cognizant of the urban environment. For example, the successful 2024 Olympic bid by the City of Paris includes the provision that certain aquatic events will take place in the Seine River. Mitigation efforts by the oversight entities therefore need to address not only wet-weather but also dry-weather pollution in urban environments; the reduction of bacteria in treated wastewater requires implementing some form of wastewater treatment plant (WWTP) effluent disinfection. Assessing and optimizing the effectiveness of such a disinfection method implies the ability to monitor FIB concentrations throughout the wastewater treatment process.

The present study describes a new wastewater measurement protocol implemented using Fluidion® ALERT technology. The calibration and validation procedures were carried out with a wide range of wastewater samples obtained from typical modern wastewater plants. ALERT results were compared side-by-side with analyses performed by an approved third-party laboratory using the MPN microplate standard method (ISO 9308-3 1998). Samples were included from all the relevant treatment stages: primary clarification, decarbonation, nitrification, denitrification, and tertiary activated carbon. Moreover, results were presented from a full-scale WWTP chemical disinfection pilot, where Fluidion® ALERT technology was deployed operationally to provide high-frequency monitoring of the *E. coli* concentration in both untreated and disinfected effluent, thereby measuring the abatement factors under several different operating conditions.

2.2 EXPERIMENTAL DESCRIPTION
2.2.1 Description of the ALERT technology

Fluidion® ALERT technology is based on a modified real-time DST method, as implemented in automated instruments capable of performing, directly in the field, the complete protocol for bacterial quantification. ALERT instruments can automate the full range of operations: sampling, reagent mixing, incubation, real-time multispectral optical analysis (absorbance/fluorescence), turbidity correction, signal analysis, bacterial quantification, wireless data transmission, and automatic generation of notifications. ALERT technology can be deployed in multiple configurations: ALERT System for performing automated *in situ* bacterial enumeration to obtain time-series data at a target location; and ALERT LAB as a portable device for the rapid mapping of bacterial contamination at multiple sites or as a bench-top device for rapid laboratory analysis (Figure 29). Both ALERT System and ALERT LAB devices have been used in the present study.

The bioreagent used in ALERT instruments contains a mixture of a selective growth medium and 4-methylumbelliferyl-β-D-glucuronide (MUG), which can be hydrolyzed into fluorescent 4-methylumbelliferyl (MUF) by the β-glucuronidase enzyme present in *E. coli* bacteria (note: MUG is the standard substrate used in approved *E. coli* testing methods). The bioreagent used also contains ortho-nitrophenyl-β-galactoside (ONPG), another bacterial indicator metabolized by all types of coliforms in the sample and transformed into ortho-nitrophenol (ONP), resulting in the development of yellow coloration. During the selective culture step (involving incubation at 37.0°C), bacterial metabolism progressively transforms MUG into MUF, in generating a broad fluorescence when excited at 385 nm, with emission peaking at around 460 nm. The growth of non-target organisms is not promoted during this selective culture step, which makes the method highly selective to culturable *E. coli*, unlike rapid tests based solely on enzymatic activity without culture.

All ALERT instruments contain multiple individual bioreactors (six for the portable ALERT LAB, seven for the *in situ* ALERT system), each capable of independently incubating a sample and performing optical measurements with an optical sensor ring. This sensor ring contains three LEDs, arranged to excite MUF fluorescence (385-nm excitation), measure ONP absorbance (430 nm) and compensate for sample turbidity (610 nm), as well as a photodiode coupled to a low-pass optical filter that blocks the UV excitation light. The fluorescence signal is measured at periodic intervals (every 5 min), with data transmitted in real time through the mobile phone network to a remote cloud-based data server. The resulting curve (Figure 30) consists of an initial plateau, followed by a sharp increase in fluorescence starting a few hours into the measurement. The curve is automatically analyzed by the data server to establish the fluorescence detection time, which is then used, after applying a specific calibration, to calculate the number of bacteria present in the original sample.

2.2.2 Laboratory reference method

The reference method employed for all side-by-side comparisons is the miniaturized MPN technique using a 96-well microplate (ISO 9308-3 1998). This method consists of performing multiple dilutions of the sample to be analyzed, according to a protocol that depends on the expected degree of pollution; the 96 wells, pre-loaded with the *E. coli* reagent, are then inoculated with the various sample dilutions and incubated. The number of positive wells upon each dilution is counted, and an MPN table is used to determine the bacterial concentration and corresponding 95% confidence interval. The dilution series used in this study has been adapted to wastewater analysis: six dilutions were performed (1/2, 1/20, 1/200, ½,000, 1/20,000 and 1/200,000); the resulting measurement range extended from 60 to 6.7×10^8 MPN/100 mL.

2.2.3 Study design

The first part of this study consisted of analyzing a number of samples side-by-side using both Fluidion®ALERT technology and the laboratory reference method, in order to obtain and then validate a calibration function corresponding to wastewater measurements. Samples were collected from the 'Seine Centre' WWTP located in Colombes, which treats 240,000 m^3 of wastewater daily, as generated by 900,000 residents of the Greater Paris Region. The plant's treatment process is typical of a modern WWTP: a pretreatment stage (screening, grit and oil/grease removal) is followed by primary settling or clarification (organic sludge and chemical phosphorus removal) and biofiltration treatment (decarbonation, nitrification, denitrification). After these stages, the final effluent is released into the Seine River. The Seine Centre WWTP is also equipped with a tertiary activated carbon filtration pilot for a fraction of the effluent; however, no specific disinfection process is currently applied. The treatment process is effective in removing carbon, nitrogen and phosphorus, with global efficiencies of between 70 and 94%. The nitrification step also results in significant FIB removal, with four orders-of-magnitude *E. coli* abatement (from 10^7 MPN/100 mL after the clarification stage to 10^4 MPN/100 mL in the final effluent);

applying an additional tertiary activated carbon treatment was shown to further reduce the bacterial load to below laboratory detection limits (Mailler, 2015).

To ensure a broad range of FIB concentrations, a total of 125 samples were collected at the end of five different stages of the treatment process: primary clarification (26 samples), decarbonation (24 samples), nitrification (24 samples), denitrification or regular effluent (32 samples), and tertiary activated carbon treatment or treated effluent (19 samples). The samples were collected in sterile polypropylene containers, homogenized and then separated into two parts. The first part was analyzed immediately using the ALERT LAB instrument, while the second was treated with sodium thiosulfate (20 mg/L) and sent for same-day delivery and analysis at an accredited laboratory for the reference *E. coli* analysis.

During the initial phase, 49 samples were analyzed using the standard surface water protocol and calibration that had previously been developed for river water (Angelescu *et al.*, 2018a), in order to test its applicability to wastewater (test phase). Following several initial measurement anomalies (described below), it was decided to adapt the protocol by adding a sample dilution step (1/4 dilution in deionized water). The revised protocol, applied to 41 samples, eliminated the observed anomalies and resulted in establishing a new calibration specifically adapted to wastewater analysis (calibration phase). This new calibration was then validated on 35 additional samples (validation phase).

The second part of this study consisted of operationally deploying ALERT instrumentation at the Valenton WWTP during the full-scale disinfection trials. PFA was injected into the plant's effluent stream, which then traveled through a subsurface passage to the discharge point in the Seine River (Figure 28). The effluent transit time between the injection and discharge points was approximately 10 min, hence representing the disinfection product contact time. Two ALERT system instruments were used to monitor *in situ* the bacterial concentration in near real-time of both the pre-disinfection (at the plant location, Figure 28 – n°1) and post-disinfection (at the effluent discharge location, Figure 28 – n°2) to determine *E. coli* abatement during successive 24-hour periods. Another three samples were collected manually at both locations at the beginning and end of the 24-hour period, for purposes of laboratory verification.

All data from the ALERT instrumentation were sent in real time to a central data server, which ran the detection algorithm and output the resulting *E. coli* quantification. The automated ALERT data and laboratory reference data were then centralized and analyzed using Microsoft Excel, as well as custom data processing scripts written in Python language.

2.3 RESULTS OF A SIDE-BY-SIDE LABORATORY COMPARISON

The initial test phase of the study involved analyzing 49 samples from all treatment stages using the previously developed surface water protocol (Angelescu *et al.*,

2018a). On the more heavily concentrated samples (primary clarification and decarbonation stages), the fluorescence signal curves often showed aberrant behavior, with the fluorescence signal increasing linearly immediately after sample collection and without the typical plateau observed in Figure 30. This observation is typical of samples that contain very high concentrations of free MUF enzymes not contained within *E. coli* cells, even before the bacterial metabolism starts its production. Such samples can lead to major quantification errors; therefore, it was decided to modify the protocol by applying a 1/4 dilution step to the wastewater samples, thus decreasing the initial enzyme concentrations and leading to the recovery of normal signal response curves. This revised WWTP protocol has been employed throughout the remainder of the study.

During the calibration phase, 41 samples from all treatment stages were analyzed side-by-side by both ALERT technology and the laboratory reference method. Detection was performed automatically; fluorescence detection times t_{fluo} for the samples analyzed herein ranging from 4.25 to 8.50 hours, with shorter times corresponding to higher concentrations. The laboratory concentration C was plotted against t_{fluo} in a log-linear graph; a linear correlation was observed, yet the slope of the previous surface water calibration was found to be different.

A new calibration specific to wastewater samples (WWTP calibration) was therefore established between C (measured in *E. coli*/100 mL) and t_{fluo} (measured in hours), that is,: $\log_{10}(C) = a \times t_{fluo} + b$, with a and b being derived by least-squares regression (Figure 31, left). The new calibration (dashed line), as well as the previous calibration (solid gray line), are also shown on the plot.

During the validation phase, 35 additional samples were analyzed using the newly developed WWTP protocol/calibration; results were compared against the laboratory reference method.

Figure 31 (right) displays all data points from the study, separated by their respective phase (test, calibration, validation), with all points corresponding to aberrant curves from the test phase being removed. The error bars correspond to the statistical 95% confidence intervals of the MPN measurements, as communicated by the laboratory. It can be confirmed that all results closely match the calibration curve, thus indicating good correspondence over more than five orders of magnitude, with a global Pearson correlation coefficient at 88.9%, as calculated on the actual measurement values and not on the base-10 logarithm values. 46% of the ALERT results differ by less than a factor of two from the laboratory reference measurement, while 93% differ by less than a factor of five.

The histogram shown in the inset in Figure 31 is symmetrical and closely approximated by a log-normal distribution centered at 0 (red line), implying that the Fluidion® ALERT measurement method does not introduce any systematic errors after calibration. The mean standard deviation of the $\log_{10}$ difference distribution over the full measurement range is: $\sigma_{ALERT-LAB} = 0.387 \log_{10}$ units, which combines the uncertainty of the laboratory reference method with that of the ALERT WWTP method (acting as independent normal variables):

$\sigma_{ALERT-LAB} = \sqrt{\sigma^2_{ALERT} + \sigma^2_{LAB}}$. The mean statistical standard deviation of the reference method over the full range of concentrations can be calculated from the 95% confidence intervals provided by the laboratory, thus yielding: $CI = 0.304$ $\log_{10}$ units. The mean statistical laboratory standard deviation is then obtained as: $\sigma_{LAB} = CI/1.96 = 0.156$ $\log_{10}$ units (assuming a normal distribution). We can therefore infer an upper limit for the mean standard deviation of the ALERT WWTP measurement: $\sigma_{ALERT} = \sqrt{\sigma^2_{ALERT-LAB} - \sigma^2_{LAB}} = 0.354$ $\log_{10}$ units. Let us note that this is likely to be an overestimation of the actual measurement uncertainty since the actual laboratory standard deviation is significantly higher than the purely statistical confidence interval related to the MPN method, which does account for uncertainties due to sample degradation in transport or to homogenization errors. As estimated elsewhere (Angelescu *et al.*, 2018a), the actual laboratory uncertainty obtained by comparing measurement duplicates may be significantly higher, which would lead to a lower value for σ_{ALERT}. Unfortunately, duplicate measurements were not available in the context of this study.

It is also important to recognize that a notable difference exists between the ALERT method and MPN methods, in that the former measures the actual activity of all viable and culturable bacteria in the sample volume (without requiring sample homogenization), whereas the latter divides the sample volume into a finite number of aliquots (microplate wells), in assessing the number of aliquots registering positive for the bacteria of interest. While a fully homogenized sample should, in theory, lead to equivalent results using either method, in reality the effectiveness of laboratory homogenization may depend on the nature of the contamination and can be imperfect, thus leading to the presence of particles aggregating large number of bacteria in some aliquots, yet which are still counted as single positives. ALERT technology, on the other hand, measures the full viable and culturable bacterial load, which may be a more relevant parameter for the potential health impacts of contaminated water. As shown in this study, even though ALERT measurements correlate very well with laboratory MPN results in the vast majority of cases, differences may appear for samples containing a large fraction of bacteria aggregated on particles, which may be underestimated by the MPN method.

Figure 32 below further displays the samples analyzed during the WWTP protocol calibration and validation phases, color-coded by the treatment stage where each sample was collected (as indicated in the legend). The right-hand panels focus on results obtained per individual treatment stage. It can be observed that the newly developed WWTP protocol for ALERT technology is able to accurately measure all the types of wastewater samples tested, regardless of their origin or concentration.

One of the ultimate objectives of this work is to develop a rapid method for assessing the impact of WWTP effluent disposal on surface water quality in view of recreational activities. The previous analysis was therefore reinforced by applying standardized guidelines (US EPA 2014) for confirming the applicability of site-specific novel *E. coli* measurement methods in assessing recreational water quality. The US EPA site-specific criteria rely on calculating the association between two methods, by applying acceptability thresholds to both the index of agreement *IA* and R^2 value, calculated using the base-10 logarithm of the measurement values from each method. The US EPA criteria consider that agreement between the two methods is sufficient when $IA \geq 0.7$; if this condition is not satisfied, then the R^2 value allows estimating how well the two methods are correlated. If $R^2 > 0.6$, then the alternative method can be applied, but with new numerical limits that need to be derived. When applied to all data obtained using the WWTP protocol, we calculate $IA = 0.98$ and $R^2 = 0.92$, which lie above the thresholds required by the standardized guidelines (0.7 and 0.6), thus indicating excellent agreement between the two methods as per the US EPA site-specific criteria for recreational water quality.

2.4 COMPARISON OF *IN SITU* RESULTS ON WWTP EFFLUENT DISINFECTION

Results from the second part of the study, consisting of ALERT instrumentation deployment at the Valenton WWTP (France) within an operational environment, have enabled measuring the *in situ* efficacy in a full-scale chemical disinfection pilot using PFA. Seven measurements were performed by each of the pre- and post-disinfection ALERT systems during the disinfectant injection experiment, at 4-hour intervals. A 10-minute delay was introduced between the start of the pre- and post-disinfection measurements, in order to account for the transit time of the effluent between the two installation sites. Three samples were also collected by hand from each location, at the beginning and end of the 24-hour measurement period, and then sent to a certified laboratory to measure the *E. coli* content using the reference method. The results are presented graphically as a time series (Figure 33) and summarized in Table 12.

The ALERT and laboratory reference measurements produced similar measured concentrations both pre- and post-disinfection, with abatement factors of 3.4 and 2.7 $\log_{10}$ units as measured by the ALERT system and laboratory, respectively.

Key points

- ALERT technology was tested on various wastewater matrices and a new calibration was developed and validated in side-by-side comparisons with the laboratory reference method.
- The ALERT method was shown to accurately determine *E. coli* concentrations over more than five orders of magnitude, corresponding to all wastewater treatment stages: primary clarification, decarbonation,

nitrification, denitrification, and tertiary treatment. The ALERT WWTP protocol required only a single initial dilution at 1/4 (as compared to six dilutions needed for the reference method).

- Total measurement time for the samples analyzed in this study ranged from 4.25 to 8.50 hours, with shorter times corresponding to higher bacterial concentrations (as compared to the 24–72 hours required to obtain the reference method results).
- Following laboratory validation, ALERT technology was integrated into an industrial wastewater chemical disinfection pilot to monitor the operational performance and determine the abatement factor obtained through different disinfectant exposures, both *in situ* and in near real-time. Significant improvements in reliability and logistical complexity were demonstrated through deployment of automatic ALERT technology, leading to faster results while fully eliminating the need for laboratory analysis and avoiding sample degradation due to transportation or human error.

Effectiveness of PFA disinfection implemented at full scale (Seine Valenton WWTP)

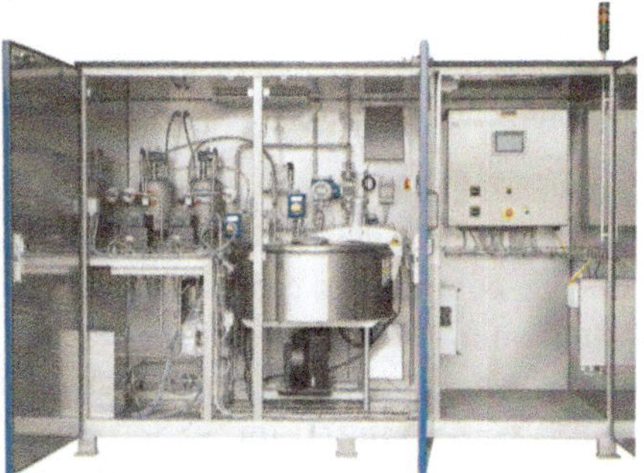

© KEMIRA

3.1 INTRODUCTION

The industrial-scale trials performed at the Seine Valenton (SEV) wastewater treatment plant (WWTP) by SIAAP in 2018 were aimed at confirming the effectiveness of PFA on fecal bacteria and identifying potential optimization levers. These trials were conducted between August and October 2018 by disinfecting the totality of the WWTP discharge (nominal capacity: 550,000 m^3/day) using PFA on a week-on/week-off basis. A complete description of these industrial-scale trials is given in Part 2, Chapter 1. This third chapter will present the full set of results on PFA disinfection effectiveness generated from these trials. First, the fecal bacteria removal effectiveness will be discussed in

doi: 10.2166/9781789062106_0075

relation to the $C \times t$ (concentration $\times$ time) of PFA applied, along with the variability induced by daily variations in the SEV WWTP discharge and the impact of PFA on other pathogens (spores of sulfite-reducing anaerobes and F-specific RNA phages). Second, the impact of SEV WWTP discharge quality variations on fecal bacteria removal will be studied with a focus on the effect of a degradation in SEV WWTP operations and the potential for results to be normalized by initial water quality. Third, the prediction of *E. coli* and intestinal enterococci removal in SEV WWTP discharge by a statistical analysis of the industrial-scale results will be investigated. Fourth and last, the impact of PFA disinfection on the discharge quality will be quantified in terms of both conventional quality parameters and disinfection byproduct formation.

3.2 EFFECTIVENESS OF PFA DISINFECTION APPLIED TO SEV WWTP DISCHARGE

3.2.1 Fecal bacteria removal achieved during the industrial-scale trials at SEV WWTP

The overall quality and variability of the SEV WWTP discharges were in the normal range and are given in Table 13, along with the quality after disinfection. The reducing compounds capable of reacting with PFA (Ragazzo *et al.*, 2013), such as ammonia, nitrite and COD, were of very limited quantity in this effluent. In comparison, the average concentrations in samples used for laboratory-scale trials were: 8.7 ± 5.5 mg/L for TSS, 5.8 ± 0.8 mgC/L for DOC, 23 ± 8 mgO$_2$/L for COD, 2.0 ± 1.2 mgO$_2$/L for BOD$_5$, 1.1 ± 0.7 mgN/L for TKN, and 1.2 ± 0.7 mgP/L for TP. Similarly, the *E. coli* and intestinal enterococci concentrations in the samples were similar to what is normally encountered in this type of water, with median $\pm$ min-max log concentrations of: 4.27 ± 3.84–5.55 and 3.56 ± 3.03–4.93 MPN/100 mL, respectively, in the industrial-scale test samples (Rocher & Azimi, 2016). In addition, these results are highly comparable to log concentrations measured during laboratory-scale trials, that is,: 4.16 ± 3.23–5.83 (Mann–Whitney test, p-value $= 0.180$) and 3.67 ± 2.85–4.80 MPN/100 mL (Mann–Whitney test, p-value $= 0.533$), respectively, for *E. coli* and intestinal enterococci. The variability in water quality originates from different SEV WWTP operating conditions over the period (internal bypass, dry and rainy weather, etc.).

Figure 34 presents the results of the 43 campaigns performed within the PFA industrial-scale disinfection unit treating the SEV WWTP discharge. It represents the median concentration of *E. coli* or intestinal enterococci, with the 1st and 3rd quartile concentrations observed as error bars (upper part of the figure) and the average logarithmic removal of both fecal bacteria with the standard deviation shown as error bars (lower part). The quality limits (900 NPP/100 mL for *E. coli* and 330 NPP/100 mL for intestinal enterococci) to respect the European recreational bathing regulation (*Directive 2006/7/EC of 15 February 2006 concerning the management of bathing water quality and repealing Directive 76/160/EEC, 2006*) are provided as indications. During these campaigns, the actual PFA dose applied varied between 0.61 and 2.48 ppm of PFA (median value: 1.20 ppm), while depending on the SEV WWTP discharge flow, the hydraulic retention time in the discharge tunnel was 19.4 and 56.1 min (median value: 29.6 min), yielding C × t values ranging from 16 to 74 ppm. min (median value: 32 ppm.min). Four classes of C × t are depicted in Figure 34: 15–30, 30–50, 50–70, and 70–75 ppm.min. The water temperature is typical for the summer period, with an average and standard deviation of 23 ± 1°C.

Similarly to the laboratory-scale trials, Figure 34 highlights the strong impact of PFA C × t applied to both the logarithmic removals and residual concentration of *E. coli* and intestinal enterococci. The logarithmic removal is stable around 2 log, between 15–30 and 30–50 ppm.min, but increases to 2.5 log above 50 ppm.min. Moreover, the residual concentrations of both pathogens decrease when increasing the PFA C × t, until reaching the limit of quantification at >70 ppm. min. The PFA C × t required to reach a sufficient quality for bathing is 10–30 ppm.min, while it rises to 50 ppm.min in order to reach residual concentrations below 100 MPN/100 mL.

The logarithmic removals of *E. coli* and intestinal enterococci are similar at a comparable PFA C × t; this similarity, for example, becomes statistically significant at a C × t of 30–50 ppm.min (Student test, *p*-value = 0.333). At the laboratory scale (Part 1, Chapter 1), the removal of intestinal enterococci was systematically 1 log lower than that of *E. coli*. Laboratory-scale removals were thus slightly higher than industrial-scale removals for *E. coli*, whereas they were comparable for intestinal enterococci. This finding proves to be significant for a C × t of 10–30 ppm.min since the average removal of *E. coli* is 2.3 ± 0.6 at the laboratory scale vs. 1.9 ± 0.7 at the industrial scale (Student test, *p*-value = 0.033); for the same C × t, this value is 1.7 ± 0.6 at the laboratory scale and 1.8 ± 0.7 at the industrial scale for intestinal enterococci (Student test, *p*-value = 0.996). Given that the exposure time was longer during industrial-scale testing (median value: 29.6 min) than the laboratory-scale trials (10 min), a comparable C × t is obtained for a lower PFA dose at the industrial scale. Even though C × t does drive fecal bacteria disinfection by PFA, this result would indicate that the PFA dose also exerts a major effect, in that an overly low dose can limit effectiveness.

3.2.2 Daily variations in fecal bacteria concentrations and impact on the PFA disinfection effectiveness

Figure 35 displays the elimination performance of *E. coli*; Figure 35a shows the residual concentrations obtained after disinfection, over a 24-hour cycle and for various PFA doses, while Figure 35b illustrates the removal rate calculated during this cycle. In Figure 35c, the residual concentrations of *E. coli* are shown vs. the various C × t values. The correlation between removal rate and concentration level of the *E. coli* to be disinfected is shown in Figure 35d.

3.2.2.1 Evaluation of hourly PFA disinfection performance

The residual *E. coli* concentrations were on the whole less than 400 MPN/100 mL and less than 4 MPN/100 mL (limit of quantification) for the 2-ppm treatment dose (Figure 35a). Let us also note that these residual concentrations are low and less than 100 MPN/100 mL between 8pm and 4am (illustrated by a blue band in Figure 35a and b). The concentration levels of *E. coli* are nearly systematically below 900 MPN/100 mL (the bathing threshold), except in one case for which the concentrations of *E. coli* to be disinfected were very high (3.8×10^5 MPN/100 mL). Regarding the reductions in *E. coli*, these varied between 1.6 and 3.8 log according to Figure 35b, the maximum of which is observed at a treatment dose of 2 ppm. These removal rates are logically greater for the highest doses, but do remain high overall at night, starting at the 0.8 ppm dose rate. The low residuals observed during the night suggest this trend is being induced by other parameters.

Although these results are consistent with the monitoring carried out by the reference method with the microplate (see previous section), significantly higher reductions have been recorded here (a maximum of 3.8 log vs. 3.1 log for the reference method). This difference in removal rate is not due to the performance of the analytical methods but rather to the time elapsed between sampling and analysis, which induces the matrix change (bacterial mortality). In fact, this time is zero for the ALERT system, which carries out the sampling and analysis independently, while the reference method requires 24 hours between sampling and laboratory analysis.

3.2.2.2 Focus on key operating parameters

The daily monitoring of disinfection performance has enabled identifying two key parameters to be taken into account during operations: C × t, and the levels of *E. coli* concentrations to be disinfected. Figure 35c shows that the residual concentrations of *E. coli* after disinfection are clearly correlated with the C × t parameter. Independently of the treatment doses, the higher the C × t the greater the decrease in residual *E. coli* concentrations. These low residuals were systematically observed between 8pm and 4am (Figure 35a) when a lower water flow was considered, resulting in a higher C × t value. Above a C × t of 40 ppm.min, the residual concentrations are very low or even zero. Overall, a C × t of 20 ppm.mm seems sufficient to maintain concentrations below the swimming limit of 900 MPN/100 mL, except for a single peak at which a residual of 1175 MPN/100 mL in *E. coli* was obtained for a C × t of 22.2 ppm.min (Figure 35c). The variability in *E. coli* residuals observed for the same C × t value can be explained by the concentration levels of *E. coli* to be disinfected. The higher these levels, the more elevated becomes the removal rate. Figure 35d clearly shows this relationship by setting the treatment dose at 0.8 ppm and a C × t ranging from 18 to 26 ppm.min.

In general, close monitoring of the dynamics of wastewater disinfection has made it possible to highlight the importance of taking the C × t parameter into account during industrial operations. However, it must be adapted to both the effluent flow rates and bacteria concentration levels to be disinfected.

3.2.3 Impact of PFA on other types of pathogens

Results obtained during the five sampling campaigns, in which additional pathogens were measured, are given in Table 14. Reductions in *E. coli* and intestinal enterococci are provided to control the representativeness of these campaigns, along with reductions observed for spores of sulfite-reducing anaerobes and F-specific RNA phages.

The five campaigns considered herein are representative of the full set of industrial-scale trials since the PFA dose applied covers the complete range tested (0.8–2.2 ppm), as well as the C × t (22.1–73.8 ppm.min). In addition, the TSS concentrations measured were low, except during the September 25, 2018 campaign, which represented the nominal operations at SEV WWTP. This set-up also led to fecal bacteria log concentrations of 4.0–4.7 for *E. coli* and 3.2–4.1 for intestinal enterococci. The log removals observed for both pathogens confirm the representativeness of these campaigns, with removals of 1.4–2.3 log for *E. coli* and 1.1–2.5 log for intestinal enterococci, which is consistent with the results shown in Figure 35. The impact of TSS presence in SEV WWTP discharge is observable on fecal bacteria, with higher initial concentrations for both bacteria, thus leading to higher residual concentrations after disinfection. This effect is also noticeable on both the other pathogens, as the SSR concentration is multiplied by six while the phages are only detected during this campaign.

Regarding SSR reduction by PFA, it can be concluded that such levels of PFA dose and C × t have a limited impact on this virus given that for two of the campaigns, a very low reduction of around 0.1 log is observed while for the three other campaigns no reduction was observed. Regarding F-specific RNA phages, it is difficult to conclude based on these results since this pathogen was quantified in just one campaign, at a low concentration of 170 PFU/100 mL, and totally eliminated after disinfection. These industrial-scale results are very consistent with those obtained at the laboratory scale (Part 1, Chapter 1).

3.2.4 Influence of WWTP effluent quality on fecal bacteria removal

3.2.4.1 Impact of SEV WWTP treatment degradation on disinfection effectiveness

Figure 36 presents a comparison of campaigns performed at the industrial scale at a comparable PFA C × t of 30–50 ppm.min under both nominal ($n = 8$) and degraded ($n = 9$) SEV WWTP operations. The PFA dose injected during these campaigns was similar, 1.0 or 1.2 ppm on nominal operating days and 1.2 ppm on days with degraded operations.

The degradation of SEV treatment mainly leads to significantly higher concentrations of TSS (Mann–Whitney test, p-value $= 0.010$), which was identified as a parameter influencing PFA effectiveness, along with slightly lower conductivity (Mann–Whitney test, p-value $= 0.011$). For these campaigns, the quality of wastewater observed during degraded operations was notably different from the quality observed under nominal operations for specific parameters, such as: TSS (4.1 ± 1.4 mg/L for nominal vs. 8.1 ± 3.4 mg/L for degraded operations), conductivity (1138 ± 26 vs. 1046 ± 34), COD (19 ± 2 vs. 27 ± 3 mgO$_2$/L), BOD$_5$ (1.1 ± 0.6 vs. 3.2 ± 1.1 mgO$_2$/L), or TP (0.6 ± 0.2 vs. 1.3 ± 0.8 mgO$_2$/L). On the other hand, soluble parameters such as TKN (1.0 ± 0.2 vs. 1.6 ± 0.3 mgO$_2$/L) and DOC (6.6 ± 0.5 vs. 6.3 ± 0.4 mgC/L) remained similar. The fecal bacteria concentrations are slightly higher under degraded operations, in particular for *E. coli*, but this increase is insignificant for both *E. coli* (Student test, p-value $= 0.053$) and intestinal enterococci (Student test, p-value $= 0.130$).

Regarding *E. coli*, no significant difference exists in the logarithmic removal value (Student test, p-value $= 0.972$) between the nominal and degraded WWTP operations. The average removals and standard deviations are: 2.07 ± 0.48 under nominal operations, and 2.06 ± 0.33 under degraded operations. In contrast, the degradation of SEV WWTP discharge quality, leading to higher TSS concentrations and lower conductivity, results in a significant decrease (1 log) of the intestinal enterococci removal (Student test, p-value $= 0.001$). These average removals and standard deviations are: 2.35 ± 0.30 under nominal operations, and 1.41 ± 0.59 under degraded operations. Since the initial intestinal enterococci concentrations are similar in both the nominal and degraded operations campaigns, higher residual concentrations were identified after disinfection, thus demonstrating the negative effect of TSS and/or conductivity on PFA disinfection effectiveness for this pathogen. This conclusion is consistent with laboratory-scale observations (Part 1, Chapter 1), even though the TSS effect was observable at higher TSS concentrations; it is also consistent with recent industrial-scale trials in Biarritz, which demonstrated a correlation between PFA disinfection effectiveness and both TSS content and water conductivity (Part 4, Chapter 1) (Pigot *et al.*, 2019). The particulate matter and conductivity should be monitored in-line in PFA disinfection units and moreover used as a proxy to modulate the injected PFA dose.

3.2.4.2 Normalization of PFA effectiveness to the SEV WWTP discharge quality

In Part 1, Chapter 1, it was demonstrated with laboratory-scale results that normalization of the PFA $C \times t$ applied by initial water quality parameters offered promise. In addition, industrial-scale trials displayed the negative effect of TSS on intestinal enterococci removal. Consequently, similar normalizations were carried out with industrial-scale results and have been plotted alongside laboratory-scale results in Figure 37. This figure presents the residual concentrations of both types of bacteria after disinfection vs. the normalized PFA $C \times t$. This normalization was performed by dividing the $C \times t$ applied by both initial bacterial concentrations, TSS or COD.

Figure 37 confirms with industrial-scale results the trends observed at the laboratory scale. Both *E. coli* and intestinal enterococci residual concentrations are significantly correlated with $C \times t$ normalized by the initial bacterial concentration (Spearman test, $r = -0.557$ and p-value $= 0.0002$ for *E. coli*, and $r = -0.353$ and p-value $= 0.024$ for intestinal enterococci), TSS (Spearman test, $r = -0.588$ and p-value $= 5 \times 10^{-5}$ for *E. coli*, and $r = -0.603$ and p-value $= 3 \times 10^{-5}$ for intestinal enterococci) or COD (Pearson test, $r = -0.618$ and p-value $= 2 \times 10^{-5}$ for *E. coli*, and $r = -0.537$ and p-value $= 0.0003$ for intestinal enterococci). Power law-type relationships can be determined in order to predict residual bacterial concentrations from $C \times t$ and the initial value of either bacteria, TSS or COD. This approach demonstrates that in-line measurements of quality parameters like TSS and DCO are consistent when regulating PFA injection, even if the approach does not totally eliminate the variability in results.

It is also interesting to observe that for intestinal enterococci, industrial-scale results are highly comparable to laboratory-scale results since both point clouds follow very similar power laws, regardless of the normalization. In contrast, an offset of 0.5–1.0 log is observed with *E. coli* industrial-scale results compared to the corresponding laboratory-scale results, thus leading to different power laws. The fact that the initial concentrations of fecal bacteria and conventional quality parameters are similar in both groups of water samples is explained by the slightly lower removals obtained at the industrial scale for this pathogen, originating from a dose limitation, than at the laboratory scale.

3.3 MATHEMATICAL CORRELATION

To further investigate the correlations observed in Section 2.1 between the disinfection efficiency observed on-site at the SEV WWTP discharge and the various operating parameters, a modeling study has been conducted. In the past, several works have proposed models of varying complexity in order to simulate the disinfection of fecal indicators using different disinfectants (Fernando, 2009; Flores *et al.*, 2016; Hassen *et al.*, 2000). Such models typically involve application of the Chick–Watson or Hom Equation (Azzellino *et al.*, 2011), which can be summarized by Equation (3.1) below:

$$\ln\frac{N_t}{N_0} = -kC^n t^m \tag{3.1}$$

where N_0 and N_t are respectively the initial and current fecal indicator concentrations (typically expressed in CFU or MPN/100 mL), k is the disinfection kinetic constant ($L^3 M^{-1} T^{-1}$), C the disinfectant concentration (ML^{-3}), and t is the time elapsed since the injection of C (T). n and m are fitting constants that account for the varying nature of disinfectants potentially used as well as other external conditions not directly taken into consideration in Equation (3.1), with m being set at 1 for the Chick–Watson equation. The initial value of C is usually considered to be constant as a means of model simplification. Other slightly more complex equations can also be used, such as Collins–Selleck. Regardless, these kinetics equations must be coupled with a description of the reactor hydraulics to properly simulate disinfection.

While reportedly appropriate for disinfectants such as chlorine or ozone, this approach is less relevant when modeling the performance of peracids (Azzellino *et al.*, 2011). Since it is known that peracid concentration decreases upon injection and slowly during disinfection, the constant Ct approach of the basic Chick–Watson equation is often considered as insufficient to properly describe the varying rates of fecal indicator removal. An integral Ct approach has thus been proposed instead (Santoro *et al.*, 2007, 2015). The disinfection kinetics of peracetic acid (PAA) have been more widely studied in the literature than those of performic acid (PFA). Various authors have proposed using a second equation to describe the variation of C over time. This variation includes an instantaneous decrease due to an immediate initial demand followed by a first-order decay (Domínguez Henao *et al.*, 2018; Murray *et al.*, 2016). In particular, Domínguez Henao *et al.*, (2018) investigated the impact of wastewater composition on the kinetics of PAA decay and *E. coli* disinfection at the laboratory scale; they observed that both the initial demand and decay rate of PAA increased to varying extents once a threshold TSS or soluble COD concentration had been reached. Likewise, the disinfection efficiency decreased at a similar equivalent PAA dose exposure when comparing samples containing, respectively, 0 and 40 mg/L of TSS.

Such modeling efforts can be beneficial not only to better assess the impacts of various factors on disinfection efficiency, but also for on-site process control. The appropriate dose of disinfectant needs to be estimated and controlled in order to maintain a sufficient level of disinfection. These efforts also help minimize both the injection cost and remaining disinfectant concentration at the plant outfall. A suitable model can thus be used either as an optimization benchmark platform or in a feedforward-type control scheme for disinfectant injection (Manoli *et al.*, 2019; Murray *et al.*, 2016).

Despite their intrinsic interest, conceptual approaches to disinfection modeling commonly found in the literature would be difficult to apply in the case study performed at the SEV WWTP. Even though PFA is not exactly the same disinfectant as PAA, it is still expected to be similarly consumed at the time of injection and during the disinfection process itself. Since PFA is more difficult to measure in samples and online than PAA, no values are available for either the laboratory experiments or the on-site case study campaigns. Likewise, the authors could not identify any modeling work regarding the modeling of PFA disinfection. No default or reference parameter values are therefore available to use with the Chick–Watson or peracid decay equations typically employed. In this work, a multiple linear regression approach has instead been used to simulate the effluent *E. coli* concentrations at the SEV WWTP; this approach takes into consideration the impact of both operation variables (wastewater flow rate, PFA dose) as well as wastewater quality at the injection point, despite the lack of PFA concentration measurements. Different criteria were used to select the optimal model input variables from a partial measurement dataset representing what could easily be measured online. The selected models were then compared using basic validation checks, plus leave-one-out and k-fold cross validation.

3.3.1 Model construction and validation

The dataset used consists of the 45 influent and effluent grab samples collected for analysis during the weeks of PFA injection, along with the six samples collected during non-injection week 42 (previously described in Part 2, Chapter 1). This dataset includes operation data at the time the grab sample was collected (wastewater flow rate, water height in the discharge channel, temperature, PFA dose, equivalent initial Ct) as well as measured concentrations. Some data are also available from more than one source. Conductivity and pH were measured on-site at the sampling point and once again after the arrival of samples at the laboratory. Temperature was also measured during sampling and by a sensor used by plant operators permanently installed in the discharge channel. Samples for which the effluent fecal indicator bacteria (FIB) concentrations were at the method quantification limit were removed. The effluent DOC concentrations were not considered since these data were unavailable for a small fraction of the dataset. Likewise, given that the purpose of this study was to estimate the effluent concentrations of each type of FIB, the measured intestinal enterococci concentrations were discarded as input variables for the *E. coli* model, and vice versa. In addition, since the correlation between influent and effluent concentrations for each measured variable was rather high (Pearson coefficient values ranging from 0.777 to 0.995), only the influent concentrations were retained as potential explanatory variables. This decision also served to decrease the number of potential explanatory variables, which was initially large compared to the number of experiments available for model fitting. The final datasets for each model thus contained 22 explanatory variables, for 48 valid measurements of EC (Table 15) and only 34 of IE. For this reason, only those results concerning the EC models will be presented herein.

A second dataset, containing the same number of valid measurements, was also prepared. For this case, only the explanatory variables considered to be easily measurable at the discharge channel intake with either sensors or *in situ* analyzers were kept. This dataset included the operational data, as well as pH, temperature, conductivity, EC concentration, DOC, turbidity and ammonia nitrogen; TSS were not included since TSS sensors actually often measure turbidity, which has already been included. Likewise, COD and BOD were excluded since they often constitute a composite estimation based on turbidity (particulate fraction) and UV-254 (soluble fraction). DOC was chosen instead as it tends to be estimated from sensors solely on the basis of UV-254. The EC concentration at the channel intake was also retained given that it could potentially be measured online, although probably with some delay between sampling and result. The secondary dataset thus contained 48 measurements and 14 explanatory variables, including similar measurements from the various sources detailed above. The objective of the first dataset was to obtain an 'ideal' model using every piece of data available, whereas the second one was considered in order to evaluate the possibility of using such a model for process control or a simplified on-site performance estimation.

For each dataset, a similar model building and evaluation protocol was followed. The effluent EC concentrations were natural log-transformed before use, as often takes place in FIB modeling (Eleria & Vogel, 2005). An initial Box–Cox power transform ($\lambda_{opt} = -0.16$) confirmed the choice of a logarithmic form (Herrig *et al.*, 2015). The regressions were performed according to a stepwise procedure, using the *stepwiselm* function available in Matlab® R2018b (MathWorks). Since this procedure does not necessarily guarantee finding an optimal model on the first try, the explanatory variables are identified using the Sum of Squared Error (SSE) criteria, the Aikake Information Criteria (AIC) and the Bayesian Information Criteria (BIC), by employing both a forward and backward procedure in each case. The identified models were further simplified by subsequently removing those explanatory variables for which the 95% confidence interval crossed zero and then refitting the model with the remaining variables. For each dataset, the stepwise procedures were first performed by only considering the direct effects of explanatory variables (no interactions) and including two-way interactions for all available variables during a second run (Table 15). All models identified by this stepwise procedure were verified for normal error distribution as well as for obvious trends between errors and each retained explanatory variable. The Variance Inflation Factor (VIF), normal and adjusted coefficients of determination (R^2 and adj-R^2) and the mean square error (MSE) were computed in all cases (Montgomery *et al.*, 2012).

Lastly, a cross-validation was performed for each model identified as a means of further optimal model selection, using the *crossval* function in Matlab. A leave-one-out (LOO) and a series of k-fold (for $k = 2$–10) procedures were applied. The single value of the LOO validation MSE (MSE_{val}) was considered, while the median value from a 1000 random sampling was considered for k-fold. The MSE_{val} values were compared both between models and to their construction MSE results in order to identify those models for which a large loss of accuracy was to be expected outside of the specific training dataset used.

3.3.2 Main results

The main results obtained for models found on the full dataset are summarized in Table 16. Four distinct combinations of optimal explanatory variables were found, as labeled F1 to F4. As is regularly the case, the forward stepwise selections (F1 and F2) identified models with fewer variables than the backward selections (F3 and F4). In both cases, the two identified models had a very similar structure; models F1 and F2 include a very small number of variables, based on the PFA dose, Ct, temperature and, potentially, the influent ammonia concentration.

The resulting maximum VIF value is high in both cases (23–23.1) and corresponds to the significant correlation between PFA dose and Ct value. For purposes of comparison, optimum VIF values under 5 or 10 are often reported in the literature (Herrig *et al.*, 2015; Montgomery *et al.*, 2012). The R^2 and adjusted R^2 values are relatively high in both cases, although the removal of $[NH_4^+]$ in F2 results in a decrease in these scores compared to F1. Models F3 and F4 contain more variables (eight) and are based instead on a combination of water flow rate and height, temperature, PFA dose, influent carbon concentrations and either ammonia (F3) or pH (F4). Both models exhibit very similar values of each selection criterion applied. The maximum VIF value equals 11.6, which stems from the rather strong correlation between water flow rate and channel height (Pearson correlation of 0.95). An increase in the adjusted R^2 values is observed when compared to F1 and F2, thus suggesting that the additional explanatory variables do raise model accuracy. Similarly, models F3 and F4 produce a much lower cross-validation MSE value than the 'simpler' models F1 and F2. The increase between the MSE of model construction and validation is also much lower (1.4–1.6 vs. 2.0–2.1), suggesting that the simpler models are less accurate when applied to 'new' conditions. For all these reasons, models F3 and F4 seem more appropriate for general use. F3 is, at first glance, preferable due to the slight increase in cross-validation MSE between F3 and F4; however, it could be argued that ammonia is not as easily measurable as pH.

Table 17 shows the main results obtained for models found on the 'online' dataset' without the inclusion of any variable interactions.

In this case, only two distinct models were identified. Every forward stepwise criterion identified model S1, which is exactly the same as model F1 based on the full dataset. Since the 'online' dataset contains a smaller number of potential explanatory variables while conserving every variable identified in model F1, this result is not unexpected. Moreover, every backward stepwise procedure converged on the same model (S2, Table 17). Model S2 is also quite similar to model F3 built on the full dataset, only excluding the BOD and COD concentrations, which results in a small decrease in R^2 (0.036) and adjusted R^2 (0.024) as well as an increase in the construction and cross-validation MSE results when compared to F3. The maximum VIF value is also slightly lower (10.5 vs. 11.6), which suggests that most of the key information needed to predict the effluent EC concentration could, in theory, be made available by online measurements. One important caveat of the work on this subject, however, is that the same measurements are used as potential model inputs for both the full and online datasets. The concentration values were obtained by laboratory analyses (except pH and conductivity) without any duplicate measurements using sensors. Both laboratory and *in situ* measurement performance may differ depending on many factors. Actual model performance using online sensors could thus differ from that obtained in this case. A proper data acquisition campaign would be needed in order to fully validate these findings. Once again, however, the slightly more complex model S2 can be considered as a preferable compromise between complexity and performance.

Table 18 summarizes the positive results obtained when considering two-way interactions during the model identification procedures.

Most attempts led to a very high number of variables and interactions being included, compared to the amount of data available, resulting in overfitted models with poor cross-validation MSE values. Only the forward AIC and backward SSE criteria on the 'online' dataset found models containing a reasonable number of interactions that apparently increased model accuracy when compared to previous results. Both models in Table 3.18 are similar to model S2. SI1 replaces the PFA dose by interactions of PFA dose with temperature and DOC concentration. SI2, on the other hand, replaces the water height by an interaction between height and temperature. In both cases, the temperature recorded during sampling replaces that taken by the WWTP sensor. SI1 yields an improvement in overall model performance (both construction and validation) while SI2 maintains similar construction results, albeit with an increase in cross-validation MSE values.

Figure 38 shows the model results over the entire dataset for the optimal models F3 and S2. As can be observed, restricting the available variables to 'online' measurements leads to greater model dispersion for the EC data around 5–7 log;

it also leads to underestimating the high effluent readings recorded during the week of measurement without any PFA injection. Also noticed in Figure 38 are the few concentration measurements available in the log 7–9 region (i.e., a low PFA dose and/or high water flow rate). Additional measurements in this region would be beneficial to ensure that the optimal models are suited for the whole range of possible conditions.

Although care must be taken when interpreting the physical meaning of variables identified by stepwise procedures in linear regression, a comparison of these results with factors incorporated in knowledge-based models can be useful. As highlighted in previous sections, the typical factors included in EC disinfection knowledge-based models are reaction time, initial EC concentration, and integrated Ct value, which itself is dependent on the initial acid dose and any factors affecting its speed of consumption (e.g., TSS, soluble COD, temperature). Domínguez Henao *et al.*, (2018) also considered TSS as a direct influential factor, by acting like a protection for fecal bacteria and not only through an increase in the disinfectant consumption speeds. The optimal models obtained in this study all have a similar structure; they are based on the water flow rate and discharge channel height (correlated with the hydraulic residence and hence the reaction time), the PFA dose introduced, the water temperature, and different concentrations of remaining pollution in the WWTP effluent, for the most part carbon-based. This finding is partly similar to the observations highlighted in the previous sections (Part 1, Chapter 1), directly for DOC and indirectly for TSS and the initial Ct.

The influent EC concentration, however, is missing from every identified model; this situation could be due to the relatively strong correlation in available data between the influent EC concentration and several of the explanatory variables found. Such is the case for Q (Pearson $\rho = 0.61$), H (0.58), COD (0.74) and BOD (0.84). At the effluent of a WWTP, the major increases in EC concentration are in fact typically caused by a partial bypass of the normally complete treatment lane. These bypasses are often due to either treatment process unavailability (maintenance, repair) or high flow events, such as rain periods. Partial bypasses also often result in an increase in pollution concentrations in the discharge channel. A combination of these factors could be sufficient, in this specific case, to estimate the order of magnitude of the influent EC concentration. This approach, however, also leads to multicollinearity amongst the various concentration explanatory variables. As such, the preference of COD, BOD and DOC over other variables, like TSS or turbidity, by the model selection procedure employed should not be over-interpreted. For similar reasons, the maximum VIF value for all optimal models lies above the recommended maximum of 10. As mentioned above, in most cases this situation is due to the strong correlation between water flow rate and channel height. Although additional data acquisition could potentially help reduce this VIF value, it is unlikely to be of much improvement.

3.4 INTERACTIONS BETWEEN PFA AND THE PHYSICOCHEMICAL QUALITY OF SEV WWTP DISCHARGE

3.4.1 Impact of PFA on conventional quality parameters

Together with fecal bacteria measurements, conventional quality parameters were monitored in both SEV WWTP discharge and disinfected water to determine whether PFA has a significant impact on water quality. The average concentrations and standard deviations are given in Table 13. A stability in conductivity, TSS, TKN, $N\text{-}NH_4^+$, $N\text{-}NO_3^-$, $N\text{-}NO_2^-$, TP and $P\text{-}PO_4^{3-}$, along with a sizable increase in DOC, COD and pH, can be observed. More specifically, the lack of a PFA disinfection effect on TSS (Mann–Whitney test, p-value $= 0.192$) and conductivity (Mann–Whitney test, p-value $= 0.910$) is statistically significant. The increases in DOC (Student test, p-value $= 8 \times 10^{-19}$), COD (Mann–Whitney test, p-value $= 0.001$) and pH (Mann–Whitney test, p-value $= 0.0001$) are also significant. The evolutions of DOC, COD and pH are depicted in Figure 39 for the entire industrial-scale trial period. An average increase of 1.2 mgC/L, 2 mgO_2/L and 0.19 pH unit is observed, with an average applied PFA dose of 1.2 ± 0.5 ppm. A slight change in effluent pH due to PFA was also detected by Ragazzo *et al.*, (2013) at the industrial scale even though it was attributed to acidification. Similarly, some cases of slight acidification have been reported in the literature with peracetic acid, with in most instances an insignificant evolution (Luukkonen & Pehkonen, 2017). The increase in DOC or COD was reported in the literature (Luukkonen & Pehkonen, 2017), along with a slight increase in dissolved oxygen due to peracid decomposition.

This discussion confirms, at the industrial scale, that PFA disinfection has a limited impact on conventional quality parameters in SEV WWTP discharge, with the only significant impact being on carbon due to the injection of a non-negligible quantity of carbon via PFA (Figure 39). This finding is in complete accordance with laboratory results (Part 1, Chapter 3), which measured a DOC input of 0.78 ppm C/ppm PFA injected, vs. a theoretical input of 0.79 ppm C/ppm PFA injected. The average PFA dose injected was in fact 1.2 ± 0.5 ppm, thus corresponding to a theoretical input of 1.0 ± 0.4 mgC/L. Ragazzo *et al.*, (2013) observed an increase of 0.68 mgC/L per ppm of PFA applied.

3.4.2 Production of disinfection byproducts caused by PFA

The potential for disinfection byproduct formation by PFA has been studied in detail in Part 1, Chapter 4, for both the laboratory-scale and industrial-scale trials. Table 19 displays the concentrations of adsorbable organic halogens (AOX), bromide and bromate measured in SAV WWTP discharge before and after PFA injection during the five sampling campaigns performed within the industrial-scale trials. Bromide is a precursor of both brominated oxidation byproducts and bromate.

An increase of 10–41% of AOX in SEV WWTP discharge with PFA injection is systematically observed at the industrial scale. Slight increases of AOX with PFA disinfection have also been reported in the literature (Karpova *et al.*, 2013; Luukkonen & Pehkonen, 2017), but these were notably lower when compared with other peracids or oxidants. The PFA dose and C × t applied seem to produce an effect on AOX formation since the largest AOX increase (+41%) is observed with the highest PFA dose (2.2 ppm) and applied C × t value (73.8 ppm.min). In contrast, the bromide concentration is reduced by PFA disinfection from 0 to −58%, even though the initial bromide concentrations are low (100–250 µg/L). No bromate was detected either before or after disinfection despite bromide levels being above 100 µg/L, which was defined as the concentration range where bromate can be formed by oxidation (von Gunten, 2003). This finding confirms that no bromate formation is expected with PFA at such low doses.

Key points

- Industrial-scale trials performed at the SEV WWTP confirmed the performance obtained at the laboratory scale with logarithmic removals of around 2–2.5 log for both *E. coli* and intestinal enterococci at PFA doses of 0.8–2.2 ppm and a C × t of 15–75 ppm.min.
- High-frequency measurements of *E. coli* with the ALERT apparatus highlighted the daily variability in fecal bacteria concentrations and its impact on logarithmic removal with PFA, thus leading to the conclusion that the PFA dose should be modulated between daytime and nighttime to optimize disinfection.
- PFA has a limited disinfection effectiveness on SSR and on totally inhibited F-specific RNA phages, for the only campaign during which this pathogen was detected in SEV WWTP discharge.
- The influence of SEV WWTP discharge quality variations on the PFA disinfection effectiveness observed at the laboratory scale was confirmed at the industrial scale, with better fecal bacteria removals during nominal SEV WWTP operations, corresponding to lower TSS concentrations and higher conductivity. Residual concentrations of fecal bacteria are correlated with the applied C × t per unit of bacteria in the water. Both TSS and COD serve as good proxies when employed to modulate PFA injection.
- PFA disinfection at the industrial scale led to a slight increase in DOC, COD and pH; however, no bromate formation was observed despite a decrease in bromide. A slight increase in AOX indicated the formation of disinfection byproducts, but specific results on this point (see Part 1, Chapter 4) did not display any formation of problematic byproducts.
- Two series of optimal linear regression models were developed to estimate EC concentration at the discharge channel outfall during the disinfection

trials at SEV WWTP. These optimal models contained relatively few explanatory variables and featured both high R^2 and adjusted R^2 values, thus indicating the potential interest of such a model for the online on-site estimation of disinfection efficiency. Also, the use of actual sensors to measure the concentration values identified as explanatory variables would be necessary to draw a complete conclusion on the online model's potential.

Assessment of the Eco-Toxicological Effects of Disinfection Processes on the Seine River

Chapter 1
Description of the biological models used to monitor water quality

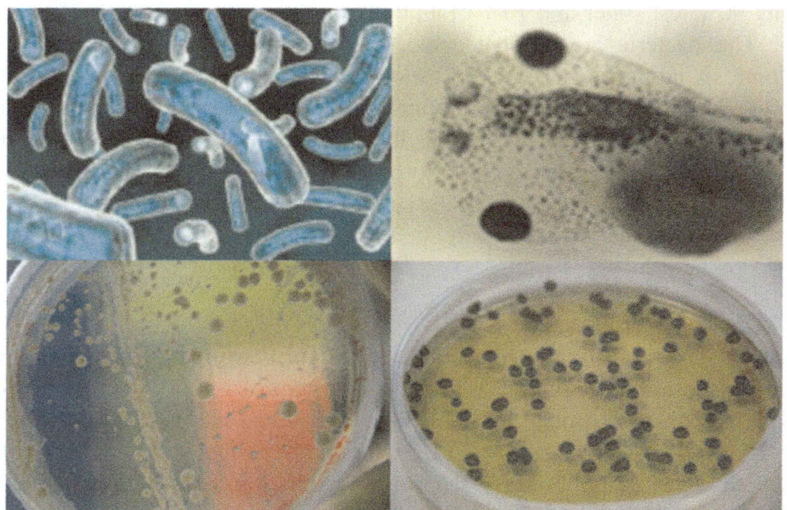

© TAME-WATER & LABORATOIRE WATCHFROG

1.1 DESCRIPTION OF THE BIOLOGICAL PANELS USED TO ASSESS GENERAL TOXICITY

The biological models used in this project to assess the general toxicity of Seine River water under the influence of chemical disinfection have been grouped in Table 20.

doi: 10.2166/9781789062106_0093

The description of each model and the representation of results obtained will be described in this section. The following proprietary conditions reflect the need for specific modifications to common microbiology protocols and thus to the experimental setting, in order to test the various water samples by directly using a 'Whole Effluent Testing' (WET) approach.

1.1.1 General toxicity in the bacterial model

E. coli bacteria are maintained in culture at 4°C in Petri dishes on a solid agar LB medium and regularly transplanted, as required, onto identical dishes. The day before the experiment, a sample is extracted and the two strains of differential sensitivities are exposed to a culture step in a liquid LB medium under gentle agitation at 37°C (12 h). The next day, the bacteria are resuspended at an optical density at 600 nm (OD600 – Victor 3 Perkin Elmer Reader) of 0.2 in a mixture containing 90% test water and 10% of a proprietary concentrated bacterial culture medium. The bacteria are incubated at 37°C under gentle agitation, and the OD600 corresponding to their multiplication is measured continuously every 40 min for 7 h. In parallel, the same experiment is performed for each strain, with a negative control consisting of double-distilled irrigation water and an ampicillin toxicity control. The response curve profiles are integrated by means of a mathematical operation that takes the entire response into account on all experimental data points, in comparing the sample curves to the controls.

1.1.2 General toxicity in the yeast model

S. cerevisiae yeast is maintained in culture at 4°C in Petri dishes on a solid agar Sabouraud medium and regularly transplanted, as required, onto identical dishes. The day before the experiment, a sample is extracted and the two strains of yeast with differential sensitivities are exposed to a culture step – 12 h for non-susceptible yeasts and 36 h for susceptible yeasts – in a liquid YPD medium under stirring at 25°C in a moisture-saturated atmosphere. On the day of the experiment, the non-susceptible yeasts are collected at a concentration of 100,000 cells/mL in a mixture containing 90% test water and 10% of a proprietary concentrated yeast culture medium; the susceptible yeasts are then repeated at 500,000 cells/mL in a proprietary medium containing 90% test water. The yeasts are incubated at 37°C under gentle agitation in a moisture-saturated atmosphere, and the OD600 corresponding to their multiplication is measured continuously (Victor 3 Perkin Elmer Reader) every hour for 7 h. In parallel, the same experiment is carried out for each strain on a negative control consisting of double-distilled irrigation water and a cadmium toxicity control. The response curve profiles are integrated by means of a mathematical operation that takes the entire response into account on all experimental data points, when comparing the sample curves to the controls.

1.1.3 General toxicity in the fungal model

The *S. tritici* strain is cultured at room temperature in Petri dishes on a solid agar PDA medium and regularly transplanted, as required, onto identical dishes. On the day of the experiment, a sample is scraped from the spore-forming areas on the jelly, and the spores are returned to suspension at 1000 cells/mL in a proprietary nutrient broth mixture containing 90% test water. Incubation is performed at room temperature (21°C). The OD600, proportional to the number of cells in solution, is measured (Victor 3 Perkin Elmer Drive) at a set time every day for 7 days. In parallel, the same experiment is performed for each strain on a negative control consisting of micropurified water and a cadmium toxicity control (3 mg/L). The response curve profiles are integrated by means of a mathematical operation that takes the entire response into account on all experimental data points, when comparing the sample curves to the controls.

1.1.4 Statement of results

The levels of proliferation or toxicity effects obtained are illustrated by a color code: green for no significant effect, red for a strong toxic effect, and dark blue for a strong proliferation effect. Each biological model has different sensitivity limits, which are taken into account in the graphic representation (Table 21). For the bacteria and yeast models, two distinct strains are used: a so-called sensitive strain, devoid of its defense systems against environmental toxins; and a natural strain (called wild-type), which does not exhibit any of these deficiencies. The sensitive model is unable to survive in the environment because it lacks all capacity to adapt and resistance to toxic stress. These conditions attest, at a very low sensitivity level, to the presence of a substance affecting living organisms. The wild-type model, commonly found in the environment, places the presence and impact of pollutants into perspective by measuring their impacts on an organism able to adapt and resist. The combination of these two information elements makes it possible to weight the interpretation of the measured signal.

Each test is based on measuring the growth of the target population in the presence of the sample to be tested. This measured growth is then compared to the reference (control) growth obtained under ideal conditions (devoid of any toxic stress). The comparison is carried out, after integrating the signal measured as a function of time, on both the growth kinetics and the final growth rate. Results are expressed as a percentage increase in the population, corresponding to the ratio of development acceleration between the stress-free control conditions and the stress test conditions.

1.2 DESCRIPTION OF THE BIOLOGICAL PANELS USED TO ASSESS ENDOCRINE DISRUPTION

1.2.1 Description of biological panels

The biological models used in this project to assess endocrine disruption in Seine River water samples following chemical disinfection have been grouped in Table 22.

Three types of endocrine disruptors are evaluated: thyroid, estrogenic, and androgenic. In order to detect these types, tests using aquatic vertebrate organisms, fish and amphibians representative of animals in the ecosystem have been used (AFNOR T90-716-1 and 2: 'Water Quality – In vivo fluorescence measurement of endocrine disrupting effects in natural water and wastewater – Part 1: Measurement of effects on the thyroid axis of amphibian embryos (Xenopus laevis) – Part 2: Measurement of estrogenic axis effects and activity of the fish embryo aromatase enzyme (*Oryzias latipes*)').

The tests using fish can detect estrogenic and androgenic disruptors, which are known to induce sex reversal in fish. Thyroid disruptors are targeted by a test using tadpoles of amphibians whose metamorphosis is affected when in contact with such compounds. More specifically, an endocrine disruption assessment is based on the use of small aquatic model organisms: tadpoles of the amphibian Xenopus for thyroid disruption, and fry of Medaka fish for estrogenic and androgenic disruption. These organisms harbor genetic markers that allow them to fluoresce when coming into contact with molecules that change how the hormonal axes function. Figure 40 illustrates the increase in fluorescence obtained when a bio-indicator tadpole comes into contact with water polluted by thyroid disruptors.

To detect all possible disruption, the samples are tested with or without co-treatment using a hormone; this step is referred to as 'stimulated' and 'non-stimulated', respectively. The larval stages used do not yet synthesize these hormones, hence this stimulation reveals certain action mechanisms of endocrine-disrupting micropollutants that cannot be detected in the absence of hormones. Tests are carried out in the laboratory as follows: larvae are exposed to plate samples (six wells) during 24 or 96 hours for fry and during 48 hours for tadpoles. The larvae are then transferred to plates (96 wells), and the fluorescence is analyzed using robotic imaging: an image is captured of each larva under specific illumination to reveal the fluorescence, and an image analysis algorithm is run to obtain fluorescence quantification. Statistical analysis is then performed to compare the fluorescence values obtained for the samples with those of the control larvae exposed to Evian water.

1.2.2 Physiological states of applied biological models

Each sample is tested in two physiological states, non-stimulated and stimulated. For the thyroid axis in the non-stimulated state, the larvae are exposed to the sample, or to the reference molecules diluted in the test medium for the control groups. In this state, it becomes possible to measure an increase in larval fluorescence induced by the test condition; this is a pro-thyroid step that activates the thyroid axis. The physiological threshold for pro-thyroid disruption (100% on the disruption scale) is demonstrated by the positive activation control, indicating the activity of the 3.25 µg/L thyroid hormone T3. This concentration corresponds to the plasma concentration of T3 hormone during tadpole metamorphosis, that is, a physiological reference for thyroid disruption as defined by OECD Test Guideline 231. Once this level of activity has been reached, it is certain that the sample has an endocrine disrupting power with deleterious effects on wildlife. In the stimulated state, the larvae are exposed to the sample or reference molecules (for control), diluted in the test medium and supplemented with T3 thyroid hormone at 3.25 µg/L. In this state, it is possible to measure an increase or decrease in larval fluorescence induced by the test condition, which indicates pro-thyroid or anti-thyroid activity, thereby activating or inhibiting the thyroid axis, respectively:

- Pro-thyroid activity (100% on the disruption scale): the threshold for pro-thyroid disruption is demonstrated by a positive control for enhancement, showing the activity of thyroid hormone T3 in the stimulated state (test medium supplemented with T3 thyroid hormone), in indicating the level of larval fluorescence after exposure to a dose of hormone that saturates the physiological response.
- Anti-thyroid activity (−100% on the disruption scale): the physiological threshold for anti-thyroid disruption is indicated by a negative control in an undisturbed state (test medium only). A tested experimental condition that induces inhibition of larvae fluorescence in a stimulated state (test medium supplemented with thyroid hormone T3) reveals anti-thyroid activity.

For the estrogenic axis in the non-stimulated state, the larvae are exposed to either the sample or reference molecules diluted in the test medium for the control groups. In this state, it is possible to measure an increase in larval fluorescence induced by the test condition. This pro-estrogenic activity activates the estrogenic axis. The physiological threshold for pro-estrogenic disruption is indicated by the fluorescence level corresponding to the dose of 64 ng/L ethinyl estradiol hormone. This dose is the lowest capable of inducing a physiological effect according to OECD test guideline 230 in Medaka fish. The physiological effect observed at this concentration is the sex reversal of a portion of the population, as identified by the presence of ovarian tissue in the testes of male fish. Once this level of activity has been reached, it is certain that the sample has endocrine disrupting power with deleterious effects on wildlife. In the stimulated state, the larvae are exposed to the sample and reference molecules, diluted in the test medium and supplemented with testosterone at 30 µg/L. In this state, it is possible to measure an increase or decrease in larval fluorescence induced by the test condition, thus indicating pro-estrogenic or anti-estrogenic activity, which serves to activate or inhibit the thyroid axis, respectively:

- Pro-estrogenic activity (100% on the disruption scale): the threshold for pro-estrogenic disruption is demonstrated by a positive control for enhancement showing the activity of 64 ng/L of stimulated ethinyl estradiol (test medium supplemented with testosterone), in indicating the level of larval fluorescence after exposure to a dose of hormone that saturates the physiological response.
- Anti-estrogenic activity (-100% on the disruption scale): the physiological threshold for anti-estrogenic disruption is demonstrated by a positive control for inhibition by treating larvae with fadrozole at 10 µg/L in a stimulated state (test medium supplemented with testosterone). The observed physiological effect is a change in the gonadosomatic ratio of male fish during the OECD test (Ankley *et al.*, 2002).

For the androgenic axis in the non-stimulated state, the larvae are exposed to the sample or reference molecules diluted in the test medium for the control groups. In this state, it is possible to measure an increase in larval fluorescence induced by the test condition. This pro-androgenic activity activates the androgenic axis. The physiological threshold for pro-androgenic disruption is demonstrated by a positive control for enhancement, in revealing the activity of 5 µg/L of 17α-methyltestosterone. This dose of hormone leads to a decrease in fertility and the presence of male secondary sexual characteristics in females (Pawlowski *et al.*, 2004). In the stimulated state, the larvae are exposed to the sample or reference molecules (for control groups), diluted in the test medium and supplemented with 5 µg/L androgen hormone 17α-methyltestosterone. In this state, it is possible to measure an increase or decrease in larval fluorescence induced by the experimental condition, thus indicating pro-androgenic or anti-androgenic activity, which serves to activate or inhibit the androgenic axis, respectively:

- Pro-androgenic activity (100% on the disruption scale): the threshold for pro-androgenic disruption is demonstrated by a positive control for enhancement, showing the 5 µg/L activity of 17α-methyltestosterone in a stimulated state (test medium supplemented with 17α-methyltestosterone), indicating the level of larval fluorescence after exposure to a dose of hormone that saturates the physiological response.
- Anti-androgenic activity (-100% on the disruption scale): the physiological threshold for anti-androgenic disruption is demonstrated by a positive control for enhancement, in showing 500 µg/L flutamide activity in a stimulated state (test medium supplemented with 17α-methyltestosterone). The physiological effect induced at a flutamide concentration of 500 µg/L is the absence of nest production due to the complete inhibition of Spiggin protein synthesis in males (Sebire *et al.*, 2008), as well as significant inhibition, in both males

and females, of Spiggin protein synthesis induced following pro-androgenic treatment (Katsiadaki *et al.*, 2006). Spiggin protein production is the physiological criterion used in the OECD GD 148 test guideline.

1.2.3 Statement of results

To specify the level of sample contamination, the larval fluorescence induced by the tested water sample is compared with that of negative (mineral water) and positive (water doped with reference hormones) controls. The fluorescence values obtained for the negative controls are consistent with the normal natural biological activity of living organisms and are not consistent with any externally induced hormonal disturbance. The fluorescence values obtained for positive controls serve to define the endocrine disruption threshold. Wherever the fluorescence intensity induced by a sample exceeds this threshold, it is likely that exposure to this sample will result in observable adverse physiological effects (Figure 41). The concentration of triiodothyronine (thyroid hormone T3) used for the positive thyroid test control is 3.25 µg/L. This concentration corresponds to the physiological load found in the plasma of tadpoles during metamorphosis, in their natural state (Leloup & Buscagalia, 1977), and is described in the literature as capable of causing observable adverse physiological effects, that is, an acceleration in the rate of Xenopus metamorphosis, and mortality if tadpoles are exposed at non-metamorphic stages (Shi *et al.*, 1998). For the positive estrogen test control, the ethinyl estradiol concentration used equals 64 ng/L, which is described in the literature as being capable of altering fish reproduction (Seki *et al.*, 2002).

Beyond illustrating results in this qualitative form in order to provide information on the ability of the sample to induce a harmful physiological effect, these results are also expressed in quantitative form. In each experiment, in conjunction with the larvae being exposed to the samples, some larvae are exposed to a range of different reference hormone concentrations that generate a 'standard' curve defining the fluorescence obtained as a function of hormone concentration. Using this 'standard curve', the fluorescence values derived for the samples can be converted into hormonal equivalents and thus establish the hormone concentration that would induce the same fluorescence variation as the sample. Use of these hormone equivalents makes it possible to compare results obtained for samples tested in different experiments.

Chapter 2

Biological models applied to the case of chemical disinfection using PFA

© SIAAP

2.1 TOXICITY ASSESSMENT AT THE LABORATORY SCALE

2.1.1 Description of the experimental approach

The general toxicity tests for PFA were carried out using spot samples of the Seine Valenton WWTP (SEV) discharge and Seine-Choisy water upstream of the SEV plant. In order to assess the impact of the disinfected PFA discharge from the

doi: 10.2166/9781789062106_0100

SEV plant on the natural environment, two key points were taken into account. The first concerns the preparation of samples to be tested, which entails reconstituting a representative sample of the natural medium in the immediate vicinity of the station discharge (Figure 42).

To this end, 'unfavorable' conditions were adopted, corresponding to a proportion (10% by volume) of water being discharged into Seine River water; this proportion was calculated based on a high summer discharge rate (90th percentile over the period 2016–2017) and a low Seine flow upstream of SEV (10th percentile over the summer period 2016–2017). For the disinfection of discharge water, a shock concentration of 2 ppm PFA was applied. Two types of samples were obtained: Seine River water containing discharge water before disinfection, and Seine water containing disinfected discharge water. For the latter, the discharge water was obviously disinfected before coming into contact with the Seine water. The second point pertains to the instability of the PFA disinfectant. It was decided to prepare the disinfectant just prior to launching the various bioassays.

2.1.2 Estimation of general toxicity

The general toxicity to bacterial, yeast and fungal models applied during the three laboratory-scale tests is depicted in Figure 43. These bioassays were conducted on two types of samples: Seine water samples containing non-disinfected discharge water (indicated in the figure as 'samples without disinfection'), and Seine water samples containing the same discharge water disinfected at a 2 ppm PFA concentration (indicated 'samples with disinfection').

The levels of proliferation or toxicity effects obtained are color-coded, ranging from green for no effect to either red for a strong toxicity effect or blue for a strong proliferation effect. Results are expressed as a percentage of population increase, corresponding to the ratio of development acceleration between stress-free control conditions and test conditions. In the remainder of this section, each test will be discussed separately before concluding on the overall toxicity.

2.1.2.1 General toxicity in the bacterial model

Results obtained from samples tested at the laboratory scale show an absence of general toxicity to bacterial models, for samples containing discharge water either before or after disinfection (Figure 43a and b). The two models tested, both wild-type and susceptible, display values of growth difference from control ranging from −12 to 7% and from −1.5 to −19.2%, respectively. As indicated in Figure 43a, these values are no different from the reference value (green color code). The bacteria do not appear to be impacted by the samples tested, even on the susceptible model without its defense systems. The presence of disinfected discharge water at the shock concentration (2 ppm PFA) is not, under these conditions, toxic to the bacterial model.

2.1.2.2 General toxicity in the yeast model

Results obtained from samples tested at the laboratory scale show a lack of general toxicity to the yeast model for the wild-type strain (Figure 43c and d). The values of growth difference from control, which range from −1 to 13.7%, reflect a similarity to the baseline state. On the other hand, for the susceptible model, a slight toxic effect is revealed in the test samples: for this strain, a growth differential of −12.7 and −32.1% with respect to the control is obtained for samples containing the disinfected discharge water. These values correspond to a zero effect for two of the three samples, but the most impactful value (−32.1%) indicates an effect described as medium. On samples containing disinfected discharge water, toxic effects are identified by a growth reduction ranging from −18.9 to −27.3% compared to the control. These toxic effects occur only on the sensitive model and remain moderate, hence qualifying on the impact scale as a 'weak but significant effect' according to the classification code. Yeasts without defense systems appear to be systematically impacted by the presence of disinfected discharge water, compared to the same samples without prior disinfection. The presence of disinfected discharge water at a shock concentration (2 ppm PFA) has no effect on the wild-type yeast model, though it does exert a small yet statistically significant impact on the susceptible model.

2.1.2.3 General toxicity in the fungal model

The results obtained on samples tested at the laboratory scale show a slight proliferation of the cell population (Figure 43e). Samples containing previously disinfected discharge water do not appear to differ in this respect from samples containing discharge water prior to disinfection. Values range from −3.8 to 101.2% for samples before the disinfection of discharge water and from 4.1 to 92.44% for samples containing disinfected discharge water. These values indicate the presence of moderate to strong proliferation effects, likely reflecting the presence of organic matter in the medium coupled with an absence of specific stressors in the test fungi physiology. These effects are observed for both types of samples regardless of the presence of disinfectant.

The presence of disinfected discharge water at a high concentration (2 ppm PFA) does not appear to alter the propensity to induce cell proliferation, as observed for Seine water samples containing discharge water prior to disinfection. The observed effects are due to the presence of factors conducive to fungal cell multiplication.

2.1.3 Estimation of endocrine disruption

2.1.3.1 Thyroid disruption

The results obtained for the three campaigns show, for all samples, no effect on the thyroid axis. Whether the Seine water is supplemented with discharge water before disinfection or supplemented with disinfected discharge water, the values lie below the thyroid test detection limit.

2.1.3.2 Estrogenic disruption

An evaluation of estrogenic disruption of the three samples tested is shown in Figure 44.

These results are expressed in both hormone equivalents and physiological effects in each state, that is, non-stimulated and stimulated. The physiological threshold (as represented by the red zone) is established by a dose of endocrine-disrupting reference hormone or molecule that induces an adverse effect at the organism level.

It should be noted that in stimulated mode, a co-treatment with 10 ng/L of testosterone is applied to the test sample (reference level in stimulated state). Treatment with 64 ng/L ethinyl estradiol (EE2) is then performed to identify the physiological threshold of pro-estrogenic disruption (positive activation control in the stimulated state) or 10 µg/L fadrozole to identify the physiological level of anti-estrogenic disruption (positive control for inhibition in a stimulated state). Testosterone treatment allows embryos to be placed under conditions where the estrogen axis is activated. Under these conditions, the test in fact also takes into consideration the effects on the aromatase enzyme. Aromatase is an essential enzyme for the endogenous production of estradiol from testosterone; it controls the fine balance between estrogen and testosterone in the body and thus participates in the sexual identity of individuals. Aromatase disruption is one of the OECD estrogenic disruption criteria (OECD Guidelines 229 and 230). The stimulated mode allows detecting compounds that alter the aromatase function (inhibition of protein synthesis, expression activator or inhibitor) as well as receptor antagonists. Results are expressed in terms of an ethinyl estradiol equivalent. This analysis indicates which concentration of ethinyl estradiol hormone yields the same hormonal potential as the sample. To complete this analysis, a dose response range of ethinyl estradiol is conducted in parallel with the test in order to model a standard curve. The fluorescence value obtained for the sample is then plotted on this curve and converted into an ethinyl estradiol equivalent. On the disruption scale, 0% denotes the total absence of endocrine activity and 100% refers to endocrine activity that reaches the physiological threshold.

These results assessed the absence of any estrogenic activity detectable in the non-stimulated mode, for all three tests. Let us note that the Seine water samples with disinfected discharge water do not activate the estrogen axis and moreover that adding disinfected discharge water does not activate it either. On the other hand, when in the stimulated mode by activating the estrogenic axis, an anti-estrogenic potential is detected. For 'control' samples, Seine water with disinfected discharge water has an estrogenic potential equivalent to -34, -46

and −86 ng EE2/L for the three tests, reaching 39, 38 and 72%, respectively, of the physiological estrogen axis inhibition threshold. These results, although quantifiable (by their difference with respect to controls), lie below the levels producing an adverse effect on development of the aquatic organisms under study. As for the Seine water samples supplemented with disinfected discharge water, no effect on the estrogenic axis is found in the stimulated mode for the first two tests, and a potential of −62 ng EE2/L exists for the third test.

When comparing samples with and without disinfected discharge water, differences appear between all three tests. For the first two tests, a reduction in the anti-estrogenic effect is observed after adding disinfected discharge water. This minimal effect on the estrogen axis after contact with disinfected discharge water means that the molecules in the medium no longer exhibit estrogenic activity, which suggests a phenomenon of cancelling or limiting the anti-estrogenic effect due to the disinfected discharge water. This cancellation/ limitation action could have taken place at three levels: hormonal conversion, protein expression, or receptor level. For the third test however, a decrease in endocrine potential is recorded: −86 to −62 ng EE2/L equivalent and 72 to 52% physiological effect impairment, respectively, between the two samples. This campaign has confirmed the presence of pro-estrogenic molecules or molecules that extinguish the anti-estrogenic signal.

Overall, the presence of disinfected discharge water at a shock concentration of 2 ppm PFA and under unfavorable hydrological conditions (i.e., high proportion of disinfected discharge water in the natural environment) does not exert a significant impact on the estrogenic axis. The values quantified at the laboratory scale never exceed the thresholds for physiological effects A potential anti-estrogenic effect limitation has been identified in the presence of disinfected discharge water.

2.1.3.3 Androgenic disruption

The assessment of androgenic disruption on the three samples tested is shown in Figure 45.

Results are obtained by comparing the sample to the physiological threshold in each physiological state, that is, non-stimulated and stimulated. This comparison is expressed as both a hormonal equivalent (Flutamide equivalent) and a percentage of the physiological effect threshold. This threshold is established by a dose of endocrine-disrupting hormone or reference molecule inducing an adverse effect at the body level.

It should be noted that in stimulated mode, a co-treatment with 5 µg/L of pro-androgenic hormone 17α-Methyltestosterone (17MT) is applied to the test sample (positive control of activation in the stimulated state). This threshold corresponds to the one induced by both a decrease in fertility and the presence of male secondary sexual characteristics in female medaka fish (Pawlowski *et al.*,

2004). A treatment with 500 µg/L of the anti-androgenic hormone, flutamide (in addition to 17α-Methyltestosterone), is then added. A threshold has thus been defined, as corresponding to the extinction of the pro-androgenic signal, beyond which an anti-androgenic disruption can be identified. This threshold induces an absence of nest production due to complete inhibition of Spiggin protein synthesis in male stickleback (Sebire *et al.*, 2008), as well as a significant inhibition of Spiggin protein synthesis induced in females and males following pro-androgenic treatment (Katsiadaki *et al.*, 2006). This flutamide concentration also leads to decreased fish fecundity along with histological alterations of ovaries and testis (Jensen *et al.*, 2004). Results are expressed as a hormonal equivalent (17MT or flutamide). This analysis indicates which concentration of the anti-androgen flutamide yields the same hormonal potential as the sample. To complete this analysis, a flutamide dose response range is conducted in parallel with the test in order to model a standard curve. The fluorescence value obtained for the sample is then plotted on this curve and converted into a hormonal equivalent. On the disruption scale, 0% denotes the total absence of endocrine activity while 100% refers to endocrine activity that reaches the physiological threshold.

These results allow us to assess the absence of any androgenic activity, in undiagnosed mode, for all three tests. Let us note that the Seine water samples with disinfected discharge water do not activate the androgenic axis and moreover that adding disinfected discharge water does not appear to activate it either. When in the stimulated mode by activating the androgenic axis, an anti-androgenic potential is detected for two of the three tests. The initial detection, which is very weak, was observed during the first test on the 'control' sample, corresponding to Seine water supplemented with disinfected discharge water at a potential of 7.1 µg/L flutamide equivalent, 1% physiological effect threshold and no disruption in the presence of disinfected discharge water. This difference in response can be attributed to the proximity of the values to the detection limits; moreover, the absence of an anti-androgenic effect due to the disinfected discharge water is to be considered. The second and higher detection level was obtained on the Seine water sample supplemented with disinfected discharge water; an anti-androgenic potential of 300 µg/L flutamide equivalent and 60% physiological threshold was achieved. A potential anti-androgenic effect was found in the presence of disinfected discharge water. For the third test, no disruptions were identified. Overall, in all three tests, no pro-androgenic activity could be identified, when in the non-stimulated mode. On the other hand, when in the stimulated mode, three distinct behaviors were observed for samples containing disinfected discharge water: total absence of effect, disappearance of an existing effect, and a significant occurrence of effect, in relation to the sample containing disinfected discharge water. In light of this erratic behavior, no systematic trend can be advanced concerning a potential anti-androgenic disruption due to the presence of the disinfectant. The differences in more or less quantified responses seem to depend on the seemingly evolving nature of the molecules present in the medium. Some molecules will tend to interact positively, negatively or inertly with the disinfected discharge water (or disinfection byproducts) with respect to the androgenic axis. Even though a 2 ppm PFA shock concentration was applied to the disinfected discharge water, the values quantified under these controlled laboratory-scale conditions have never exceeded the physiological effect thresholds.

2.2 TOXICITY ASSESSMENT AT THE INDUSTRIAL SCALE

General toxicity and endocrine disruption were measured on samples collected from the Seine River both upstream and downstream of the WWTP discharge point in order to evaluate the impact of WWTP discharge on the river with and without PFA injection. These samples were collected at two sampling points of the MeSeine platform, which oversees Seine River quality monitoring within the SIAAP Authority: Choisy-le-Roi (PK 622.440) (3) and Port à L'anglais (PK 626.152) (4). The Choisy-le-Roi sampling point is located approximately 1,500 m upstream of the WWTP discharge point, on the first bridge accessible to pedestrians. Sampling was performed once a week (on Wednesdays). It must be kept in mind that depending on weather conditions, single rainfall sewer overflows can occur between these two Seine River sampling spots, thus significantly impacting water quality. More information was given in a previous section dedicated to the sampling description (Section 2, Chapter 1).

2.2.1 Estimation of general toxicity in the Seine River

The monitoring of general toxicity in the Seine River during industrial-scale testing entailed tests on bacteria, yeasts and fungi; these were carried out both upstream and downstream of the SEV station outfall. During these tests, periods without chemical disinfection were also monitored in order to establish a baseline (prior to the start of testing) and assess any potential aquatic system disruptions during the post-disinfection period. Results obtained on general toxicity during the 10-week follow-up phase (between weeks 35 and 44) are illustrated in Figure 46.

As described above, for bacterial tests as well as yeast tests, two variants have been mobilized: a susceptible strain lacking certain defense mechanisms, and a wild-type strain without such defects. Results are expressed as a percentage of population growth, corresponding to the ratio of development acceleration between stress-free control and test conditions. A color code indicating the level of toxicity obtained is given in each figure, representing increased toxic effects from green to red.

2.2.1.1 Bacterial tests

Results obtained in the Seine River during the disinfection tests show an overall absence of general toxicity to the wild-type bacterial model, with negative responses in about 90% of cases, either upstream or downstream of the SEV station. The values range from -16 to 15% growth for the Seine upstream of SEV and from -5 to $+17\%$ growth alteration for downstream of SEV (Figure 46a), for the most part similar to the reference state. A small, but significant, effect was observed downstream of SEV during the first season, with a slightly more pronounced impact than that upstream. These effects are likely due to the rainy conditions of that particular day. In the other case, again a small, but significant, effect was observed in the Seine upstream of SEV that was no longer found downstream. Regarding the susceptible bacterial model, no significant effects were observed in the Seine upstream of SEV. On the other hand, downstream of SEV, a small significant effect was detected in half the cases (Figure 46b). These weak effects, comprising between -2 and -29% growth modulation for the Seine upstream of SEV and between -6 and -52% for the Seine downstream of SEV, do not appear to coincide with any particular periods of disinfection. We can in fact observe that the remainder of the time (50% of occurrences), no significant effects were detected between upstream and downstream of SEV. It should be noted that the observed toxic effects, though moderate, systematically correspond to rainy episodes. The well-known quality degradations associated with these contexts are visibly expressed by effluent inputs causing physiological stress on the susceptible strain. Laboratory-scale results on reconstituted samples of Seine water supplemented with 2 ppm PFA disinfected discharge water did not reveal a significant toxic effect of chemical disinfection on both models, that is, wild-type and susceptible, of bacteria. This finding corresponds perfectly with the results observed during the industrial-scale tests and would instead tend to support the external effluent input track (thunderstorm weirs, soil leaching, etc.) or a change in the discharge water matrix prior to disinfection, as observed by Ragazzo *et al.* (2017).

2.2.1.2 Yeast tests

Results obtained in the Seine during disinfection testing systematically showed no general toxicity to wild-type yeasts upstream of the SEV station, as well as downstream. The values ranged from -3 to $+28\%$ growth modulation for the upstream Seine SEV and between -6 and $+37\%$ for the downstream SEV (Figure 46c); these values are broadly similar to the baseline, except for two points (weeks 39 and 43). For the latter, a proliferation-stimulating effect, described as 'weak but significant', was observed in the Seine downstream of SEV, indicating the probable presence of organic matter conducive to developing yeasts. As regards the sensitive yeast model, a significant absence of effect was observed in the Seine upstream of SEV 70% of the time, compared to 80% of the time downstream of SEV (Figure 46d). These values ranged from +3 to −31% change in growth for the Seine upstream of SEV vs. between −28 and +11% downstream. Overall, during the 10 follow-up campaigns, only one showed stronger pressure downstream of SEV than upstream. This day was marked by especially high rainfall, which may explain the difference. Results observed during industrial-scale tests were consistent with the laboratory-scale tests for the wild-type strain (reconstituted samples from Seine water supplemented with discharge water disinfected at 2 ppm PFA), for which a lack of disinfection effect was noted. For the susceptible strain, a lack of industrial-scale effects was observed, in contrast with the laboratory tests, which revealed a slight, but systematic, effect after the disinfection step. This difference in response could be attributed to the more adverse laboratory test conditions (characterized by a larger proportion of disinfected discharge water in the reaction volume).

2.2.1.3 Fungal tests

Results obtained in the Seine during the disinfection tests show no general toxicity in the fungal model 60% of the time downstream of SEV vs. 40% of the time upstream. These values ranged from 3 to −55% growth modulation for the upstream Seine SEV and between −2 and −58% downstream (Figure 46e). Instances of proven impact are systematically assigned to the 'weak but significant' category and appear, interestingly, more frequently upstream than downstream of SEV. With the exception of week 37, for which an impact was observed downstream of SEV and not upstream, the minimal effects observed downstream of SEV were generally identical to those observed upstream. In other cases, either no effect between upstream and downstream or even a slight improvement over the upstream could be observed. The observed effects, though limited, thus seem to correspond not to the periods of disinfection, but rather to wider variations in Seine River quality and/or SEV discharge water. This conclusion becomes even more pronounced when taking into account various rainy events during the industrial-scale tests. Laboratory-scale fungal model testing on reconstituted samples of Seine water and discharge water, both before and after disinfection, showed no impact of disinfection in two of the three tests, with a stimulated response in the post-disinfection case. The only significant difference between industrial and laboratory tests stems from toxicity observations in the natural environment (overall negative values), whereas a tendency to activate proliferation is observed in the laboratory, due in all likelihood to the better controlled laboratory conditions, hence allowing for more efficient removal of external inputs influencing the biological test response.

2.2.2 Estimation of endocrine disruption in the Seine River

The monitoring of endocrine disruption in the Seine during the industrial-scale trials was focused on thyroid and estrogenic disruption; it was carried out upstream and downstream of the SEV station outfall. Estrogenic disruption results over the 10 weeks of follow-up (between weeks 35 and 44) are shown in Figure 47. As described above, two physiological states of the biological model were applied: the non-stimulated mode (Figure 47a and b), and the stimulated mode (Figure 47c and d). For each physiological condition, results were expressed in hormonal equivalents. A color code for the level of disruption found is indicated on the results, which are expressed as a percentage of physiological threshold attainment.

2.2.2.1 Thyroid disruption

The first thing to note from these results is the absence of thyroid disruption in the Seine during the entire monitoring period, whether upstream or downstream of SEV. Chemical disinfection applied during this period did not appear to disturb the natural environment as regards thyroid disruption. These observations are consistent with laboratory-scale results. In fact, the laboratory tests were carried out on a matrix of Seine water supplemented with SEV discharge water disinfected at 2 ppm PFA, in simulating the receiving environment immediately downstream of the station outfall under 'unfavorable' conditions. Of the three tests conducted with a 1-week interval, no disruption was quantified. These results are also consistent with studies carried out on the discharge water of the same station, in which the authors found no thyroid disruption (Du Pasquier *et al.*, 2018). According to the monitoring performed within the framework of *Meseine Innovation* (the R&D component of the Seine River quality monitoring network sponsored by the SIAAP), among the 13 sampling campaigns run between 2016 and 2017, effects ranged from 11 to 56% for both non-stimulated and stimulated modes, and moreover no effect was detected during the summer period (unpublished data).

2.2.2.2 Estrogenic disruption

For estrogenic disruption, effects were detected periodically throughout the industrial-scale testing campaign (Figure 47).

When in the non-stimulated mode, by placing the sample directly into contact with the biological model, the Seine upstream of SEV exhibited estrogenic activity most of the time and at a frequency equivalent to that downstream of SEV. The estrogenic activity detected in the Seine was in the range of 24–37 ng EE2/L upstream of SEV to 24–31 ng EE2/L downstream (Figure 47a). These effects remain low, reaching 38–58% of the physiological threshold (Figure 47b).

Overall, this low variability is due to measurement uncertainty since the results obtained between upstream and downstream of SEV are statistically equivalent (Mann–Whitney test, $\alpha = 0.05$, p-value $= 0.908$). The chemical disinfection during industrial-scale testing did not disrupt the receiving environment as regards estrogenic disruption in the non-stimulated mode. These observations are consistent with the laboratory results, for which no effects were detected either before or after disinfection.

When in stimulated mode, achieved by adding a hormone activating the estrogenic axis in the sample to be tested, an anti-estrogenic potential was detected 60% of the time in the Seine upstream of the SEV station vs. 20% of the time downstream. When this activity was detected in the Seine, it amounted to between -29 and -53 ng EE2/L for the upstream of SEV and between -25 and -48 ng EE2/L downstream (Figure 47c). The detected effects remain weak overall and range from 23 to 53% of the physiological level of estrogenic axis inhibition (Figure 47d). Results observed between the upstream and downstream of SEV, in contrast with the non-stimulated mode, are statistically significant (Mann–Whitney test, $\alpha = 0.05$, p-value $= 0.066$), thus indicating the presence of an effect due to SEV discharge water. Moreover, over the 10 weeks of follow-up between discharge upstream and downstream of SEV, an anti-estrogenic effect disappeared 50% of the time, broken down as follows: absence of effect 30% of the time, decrease in effect 10% of the time, and appearance of an effect 10% of the time. These erratic trends do not seem to coincide with the applied rates of PFA treatment or even with the periods of chemical disinfection, but rather are correlated with effluent quality (SEV and/or Seine discharge). These observations are consistent with laboratory-scale results; of the three laboratory tests conducted, the anti-estrogenic effect disappeared in two of them and decreased in the other.

In sum, during the industrial-scale disinfection tests, the Seine downstream of the ENS outfall exhibited weak endocrine disruption and equivalent, minimized or cancelled effects relative to those upstream. This finding, which is generally positive, shows the presence of molecules that no longer induce endocrine effects on organisms, either via the cancellation effect (pro- and anti-estrogenic compensation) or via inhibition, which could occur at the level of a hormone conversion process or at the receptor level. The exact action mechanism is normally difficult to precisely identify for endocrine disruptors, since they can carry out several different actions simultaneously (Du Pasquier *et al.*, 2015; Mengeot *et al.*, 2016), hence the value of combining several bioassays. This statement is not surprising, given that endocrine disruption of ETS discharge water has been estimated, based on the work by Du Pasquier *et al.* (2018), at a zero or low level.

Key points

Results obtained from samples tested at the laboratory scale indicate:

- An absence of general toxicity to bacterial models.
- No effect on the wild-type yeast model, although a small but statistically significant impact on the susceptible model.
- No alteration of the cell proliferation of fungi.
- Results obtained for all sampling campaigns showed no effect on the thyroid axis.
- No significant impact on the estrogenic axis, whereas a potential pro-estrogenic or anti-estrogenic effect limitation was identified in the presence of disinfected discharge water.
- No androgenic activity was identified, in either the non-stimulated or stimulated mode, with three distinct behaviors being observed: total absence of effect, disappearance of effect, and significant occurrence of effect, respectively, in relation to the sample containing disinfected discharge water.

Results obtained from samples tested at the industrial scale indicate:

- Regarding general toxicity models (bacteria, yeast, fungi), no significant effects linked to disinfection were observed in the Seine downstream of the SEV outfall.
- Results obtained for all sampling campaigns showed no effect on the thyroid axis.
- Chemical disinfection did not modify the receiving environment in terms of estrogenic disruption.

Section 4

Feedback from other Municipalities on the use of Chemical Disinfection with PFA

Chapter 1

Case of Biarritz (France)

Thierry Pigot and Thomas Paulin

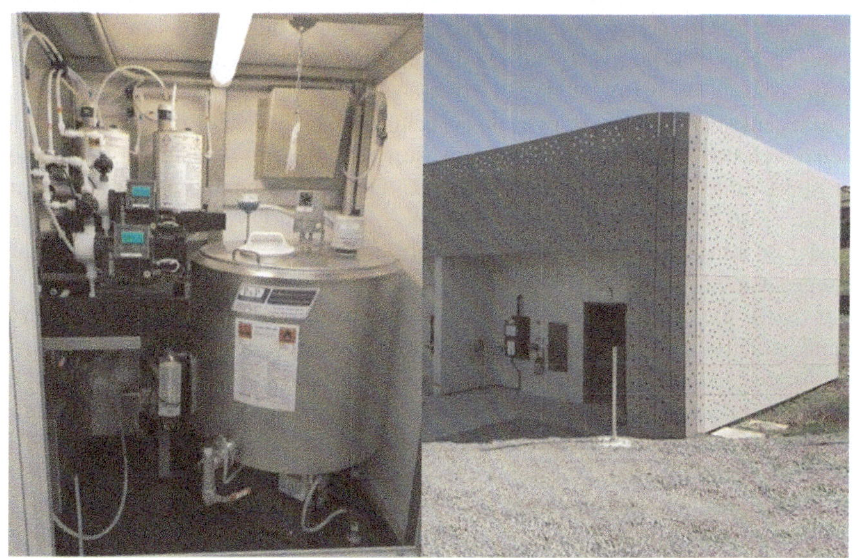

© Thierry Pigot, UPPA

1.1 INTRODUCTION

The tightening of the bacteriological quality thresholds for bathing water, following the revision of the European Directive (DCE 2006/7/EC), raises an issue for cities

doi: 10.2166/9781789062106_0115

or urban areas discharging their treated wastewater into or near bathing areas. The Urban Waste Water Treatment Directive (91/271/EEC) has imposed the implementation of a secondary treatment before discharge, but this is generally insufficient to achieve the required level of quality for bathing. The implementation of a disinfection process is then required.

There are several options available for disinfecting wastewater treatment plant (WWTP) effluents at the industrial scale: ultraviolet (UV) irradiation, chlorination, ozonation, and injection of organic peracids. UV has proven its effectiveness, despite its sensitivity to the amount of suspended solids in water. Bacterial regrowth, with the capability of damaged bacteria to adapt and repair DNA, has also been reported in the context of this technology. Chlorination is a known standard method to disinfect effluent, in particular for potable water production. However, due to the potential formation of toxic disinfection byproducts (DBP), like trihalomethanes or haloacetic acids, its use is not recommended for the disinfection of wastewater, which is rich in organic matter. Ozonation forms active species capable of disinfecting wastewater efficiently; however, bacterial regrowth and formation of DBP are also potential issues with this process.

In recent years, the use of organic peracids (peracetic acid or performic acid) has been developed for the disinfection of wastewater. Kemira (Helsinki, Finland) sells the KemConnect™, which proposes the on-site production and use of a performic acid solution (DEX solution). Compared to ozone and UV, organic peracids have lower operating and capital costs; they are also less sensitive to effluent quality changes, for example TSS and organic load.

In 2014, the city of Biarritz decided to upgrade its WWTP (Marbella) with the KemConnect™ process to disinfect the treated wastewater being discharged into the Atlantic Ocean (Bay of Biscay). Consequently, the effectiveness of this process was studied. The City of Biarritz's sanitation system, mainly unitary, collects wastewater and rainwater that are then treated at the Marbella WWTP. The treated wastewater is discharged into the ocean 800 m from the Bay of Biscay beaches via a sewer (called Milady and Marbella).

The city is coping with the potential deterioration of water quality for bathing on the bay's southern beaches, caused by a return of the outfall plume under certain oceanographic conditions and weather. This deterioration may lead to temporary closures of bathing on those beaches.

The implementation of a tertiary disinfection process in the Marbella WWTP was therefore decided in order to guarantee better bacteriological water quality at the discharge point and thus to minimize the impact on beaches during episodes of plume return.

In this chapter, the results from 18 months of full-scale operation of KemConnect™ at Marbella WWTP will be presented. The impact of KemConnect™ on the WWTP effluent quality as well as on the environment will be evaluated.

1.2 MATERIALS AND METHODS
1.2.1 Presentation of the Marbella WWTP

The Marbella WWTP has been in operation since January 1, 2004 and has a nominal capacity of 92,000 population equivalents (5500 kg BOD_5/day). Due to the specificity of the sanitation network, which is mainly unitary, the WWTP comprises a treatment line for dry weather (wastewater flow reaching the station at up to 1300 m^3/h) coupled to a treatment line for rainy weather (flow arriving at the station above 1300 m^3/h and up to 3000 m^3/h).

Beyond the rain occurring at a monthly frequency, that is when storm basins are full, the fraction of flow higher than 3000 m^3/h is discharged into the environment after a single fine screening.

Briefly, the raw effluent is first pretreated (screening, grease and sand removal) before treatment with a physicochemical lamellar decanter (Densadeg type). The dry weather system (flow rate up to 1300 m^3/h) includes an additional biological treatment by biofiltration (Biofor type biofilters), which performs nitrification and denitrification.

The treated effluents pass through a treated water storage tank and then a counting channel to be finally discharged into the ocean via an 800-meter long outfall (DN 1600) located between the Milady and Marbella beaches. Since 2014, a disinfection unit (KemConnect™) provided by Kemira offers tertiary disinfection. Performic acid (PFA) is produced *in situ* by mixing formic acid and hydrogen peroxide in the presence of a catalyst. This mixture (commercially called DEX® and containing around 12% PFA (w/w%)) is then injected into the WWTP effluent at a dose ranging from 0.8 to 1.0 ppm.

1.2.2 Sampling points and bacteria removal determination

Fecal bacteria (*E. coli* and *Faecalis enterococci*) removals were evaluated using single samples collected at two distinct points in the network. The first point lies upstream of the PFA injection point; this point allows for an estimation of the fecal bacteria load at the station outlet. The second sampling point is located downstream of the PFA injection point and constitutes the last possible sampling point before effluent discharge into the drop well and the outfall.

The effluent/PFA contact time from the injection point to the downstream sampling point was measured by means of colorimetry and estimated at 3 min, while the total contact time was estimated at 18 min until the end of the pipe and discharge into the ocean.

The upstream samples were collected in a single-use sterile plastic bottle (500 mL), and the downstream samples ($t + 3$ min, $t + 18$ min) were collected either in sterile disposable plastic bottles containing excess thiosulfate (20 mg/L) (for $t + 3$ min) or in single-use sterile plastic bottles; the reaction stopped after 15 min, with an excess of thiosulfate.

Determination of *E. coli* and *Faecalis enterococci* was performed according to standard methods (NF EN ISO 9308-3 for *E. coli*, and NF EN ISO 7899-1 for *E. faecalis*).

1.2.3 Evaluation of the environmental impact of disinfected effluent

1.2.3.1 Ecological inventory at the outfall of the WWTP discharge pipe

The objective here was to perform a qualitative and quantitative assessment of the species present at the outfall. This assessment took seasonal variations into account and was conducted over an entire season (late winter, late spring and end of summer). The observations and census protocol were carried out by two divers simultaneously and optimized in order to limit bias (size and positioning of the quadrats).

1.2.3.2 Monitoring of a bio-indicator: common mold

The biomonitoring step focused on *Mytilus edulis* (common mold), which is a bio-indicator widely used in biomonitoring; hence, extensive data are available in the literature (Bachelot, 2012; Gosling, 2003; Kerambrun *et al.*, 2012; Swiacka *et al.*, 2019).

For starters, the general condition of mussels has been evaluated using conventional morphological type descriptors based on metric and weight parameters as well as a physiological state index. To describe and compare the morphology of individuals with regard to sites and sampling dates, seven conventional biometric indices have been used (Figure 48): Length (L), Height (H), Width (W), Thickness (T), Elongation index (H/L), Compactness index (W/L), Convexity index (W/H), and Thickness index (T/L). In addition, a physiological state index was included: the Walne & Mann condition index, denoted 'IC' (Walne & Mann, 1975). This index provides information on the individual's condition (filling rate indicator) and moreover makes it possible to identify growth and reproduction anomalies; it is derived by dividing the dry mass of meat by the dry mass of shell. The higher this index value, the better the overall condition. The index varies greatly during the breeding season depending on the state of gonad development (Seed & Suchanek, 1992). Since samples are taken during the breeding period, the condition of the gonads is also observed for purposes of interpreting the results.

This approach was then completed by genetic analysis. The objective was to determine whether the individual *Mytilus edulis* present on the outfall, and continuously exposed to the PFA disinfected discharge, overexpress the genes involved in the resistance to various stresses (oxidative stress, xenobiotics, energy requirements, etc.). Twelve genes of interest were targeted for the genetic study from the sequences available in the database for *Mytilus edulis*. Individuals living on the outfall were sampled during dives conducted during the outfall ecological study (February, June and September 2015). The individuals at a reference site were collected from the Marbella dike on the days of the dives. The studied genes can be sorted into five categories, four of which are stress-response genes. The studied genes are all involved in the response to: oxidative stress (sod Cu/Zn, sod Mn, cat), mitochondrial metabolism (cox1), DNA repair mechanisms (rad51), and apoptosis (cas8, bax, bd2). The reference genes, whose expressions are constant in all cells, were studied to normalize the amount of RNA between samples. The ribosomal protein gene L7 (rpl7) is involved in protein synthesis, while the second reference gene used was the elongation factor (ef1).

1.3 EFFECTIVENESS OF FECAL BACTERIA DISINFECTION

Figure 49 displays the boxplots of the logarithmic removals obtained during the trial period until early June 2017 for both types of fecal bacteria. This removal is expressed in logarithmic units (Removal $= \log N_t/N_0$, where $N_t =$ count after treatment time t and $N_0 =$ count before treatment).

Several important observations can be drawn based on Figure 50, namely:

- For *E. coli*, with a contact time of 18 min, the average removal equals 3.3 u. log.
- For Enterococci, under these same conditions, the removal is about 1 u.log less than for *E. coli*, with an average value of 2.2 u.log. This difference is probably due to the alteration in membrane structure between the two types of bacteria, as the cell wall of Enterococci (Gram negative bacteria) is much thicker than that of *E. coli* (Gram positive).
- Significant variability exists in the removal rates for both *E. coli* and Enterococci, as demonstrated in other studies (Ragazzo *et al.*, 2013). This variability is related not only to the load of microorganisms (which can vary by a factor of 1000 from one sample to another), but also to the TSS content of the outlet water, with lower removals often being observed at high TSS concentration values. However, no simple relationship exists between fecal bacteria removal and TSS content.

In order to rationalize these experimental results, a statistical study was conducted to better define the turbidity thresholds that impact PFA effectiveness. The most suitable statistical model to account for the influence of various parameters (turbidity, TSS, temperature, conductivity, chemical oxygen demand) on a target variable (in this case the value of the bacteriological quality of water) is the conditional tree. This tool allows to show:

- which parameter exerts the primary influence on the target variable;
- what parameter value triggers a noticeable threshold effect.

Turbidity data and bacteria removal results were therefore compiled and analyzed statistically according to this methodology (using the R software rattle package, Williams, 2011); Figure 50 shows the conditional tree obtained with the bacteriological analysis value for the *E. coli* organism after 18 min of treatment with PFA. For this analysis, three target values were chosen, each corresponding to a 'bathing water' threshold: 0–500, 500–1000, >1000 (MPN/100 mL).

This conditional tree can be interpreted as follows: the most impactful parameter is turbidity, and the threshold value is 15.7 NTU, corresponding to total suspended solids of around 20 mg/L. Above this value, the proportion of analysis higher than 1000 MPN/100 mL becomes significant (about 50% of the total 14 analyses above 1000). The second parameter with a significant impact appears to be conductivity: for conductivities above 3200 µS/cm, the proportion of analyses lower than 500 MPN/100 mL is only 60%. A separation of dry weather/rainfall data does not significantly modify the analysis of the results.

All results converge to indicate that PFA effectiveness is high for turbidity below 16 NTU at a dose of 1.2 ppm (C × t of 21.6 ppm.h). Above this turbidity value, it is proposed to increase the treatment dose to 2 ppm (C × t of 36 ppm.h) and possibly to 4 ppm (C × t of 72 ppm.h) in case of extreme turbidity values.

1.4 ENVIRONMENTAL IMPACT OF THE DISINFECTED EFFLUENT

1.4.1 Ecological inventory at the outfall of the WWTP discharge pipe

The impact of WWTP discharges on benthic communities in hard substrates has been poorly documented. Several works do exist (Andral *et al.*, 2011; Cabral-Oliveira *et al.*, 2014; Terlizzi *et al.*, 2002), but the conditions are not necessarily comparable to those at the Marbella outfall. Therefore, a census of species present at the outfall has been carried out. The spatial distribution of species at the outfall (Figure 51) and its temporal evolution were monitored in 2015 and 2016, after the start of the PFA disinfection campaign in 2014. The objective of this monitoring mission was to determine whether or not ecological diversity was changing over time as a result of PFA use.

For both years monitored, the average recovery percentages show that the majority of species are *Actinothoe sphyrodeta* (47%), *bryozoans* (16%) and *Chartella papyracea* (12%).

A statistical comparison was performed in order to highlight a possible spatial distribution of colonization between the north-facing part of the outfall and the south-facing part. The result of the Wilcoxon test conducted to compare colonization between the north and south faces indicates insignificant differences (p-value $= 0.285$), which means that colonization and communities do not differ between the two areas. Consequently, the two colonization faces can be treated together without introducing bias into the analysis. The spatial distribution of benthic fauna reveals homogeneous colonization over the entire structure. In conclusion, the quadrats carried out in 2015 and 2016 do not display an effect of discards on the distribution of taxa of fixed fauna. Within these two years, a similarity is found between the dominant taxa, but with differences in average representation, particularly for *Actinothoe sphyrodeta*, *Hydrozoa* and *Sabellaria alveolata*. The presence of species considered opportunistic or, in contrast, highly sensitive to changes in environmental conditions, was not been observed during either year.

It can be concluded that the Marbella WWTP outfall is an artificial habitat where a particular biodiversity has been concentrated. This biodiversity is characteristic of hard substrates, tolerant of strong hydrodynamics, regular sedimentation and desalination (de Casamajor, 2004). In fact, the tertiary treatment does not seem to have a negative impact on the site's biodiversity.

1.4.2 Monitoring of a bio-indicator: common mold

The performance of the biological functions of bivalves is known to be strongly influenced by environmental conditions. As a result, their use as sentinel organisms and pollution controls is common (Gosling, 2003). Their mode of feeding by filtration leads to the concentration of various types of pollutants (chemical, bacteriological) and/or viruses. Thus, they are sensitive to the different stresses caused by pollution, and their growth, reproduction and longevity can be directly affected (Gosling, 2003). Among the organisms frequently used for coastal environmental biomonitoring, mussels are preferred because, as 'bioaccumulators' (i.e., capable of accumulating contaminants at levels higher than those of the environment), they allow us to study the levels and trends of chemical contamination in the aquatic environment. Accumulation performance varies with age and stage in the organism's reproductive cycle (Bachelot, 2012). Studying the biological response of organisms to pollutants combines multiple indicators, namely: physiological, biochemical, molecular and/or cellular parameters that are altered by the action of contaminants (Kerambrun *et al.*, 2012). In studies dealing with the biological effects of urban discharges in the sea, implementation strategies are generally of three types:

(1) mussel 'caging', which consists of exposing the mussels close to the outfall for a given period of time by transplanting them into cages with a control site,
(2) controlled environmental experiments in mesocosms,
(3) *in situ* sampling (De los Ríos *et al.*, 2012; Turja *et al.*, 2015).

This latter strategy has been selected for the present study. Caging tests would have been unsuccessful due to extremely unfavorable hydrodynamic conditions at the discharge point (strong swell). Moreover, mesocosm tests would require the presence of certain equipment close to the PFA production unit, which was not possible in this study.

1.4.2.1 Morphological parameters

First of all, the general condition of the mussels, as studied using conventional morphological-type descriptors based on metric and weight parameters along with and a physiological condition index, will be presented. The relationships between the various indices, which did not change substantially between 2015 and 2017, are depicted in Figure 52 via a principal components analysis (PCA). This depiction could reflect a relative stability of the environment containing the studied individuals and is consistent with their adaptation to the disinfected effluent.

Statistical tests were conducted on the condition index IC in order to compare the same sampling point in both studied years (2015 and 2017). These tests concluded an insignificant difference in this index between sampling years, meaning that the tertiary treatment implemented at the Biarritz WWTP does not degrade the condition index of mussels living on the outfall.

1.4.2.2 Genetic analysis

The morphological approach was complemented by a genetic analysis to determine if the genes involved in the resistance to various stresses (oxidative stress, xenobiotics, energy requirements, etc.) are being overexpressed. Figure 53 summarizes the results obtained on two groups of genes in 2015 and 2017: oxidative stress genes and DNA repair. For a given site and date, inter-individual variability can be high (i.e., high 'error' bars), which has also been verified herein. It is therefore generally accepted that a minimum factor of 2 for induction or less than 0.5 for repression must be achieved in order to dismiss physiological variations in expression levels between individuals (red lines in Figure 53). For a given site, the temporal variation can be extremely large and the expression levels of a gene can differ considerably depending on the sampling period. As such, outfall/reference site comparisons should only be conducted on the same date.

In considering the above remarks, no significant differences exist in terms of oxidative stress genes and DNA repair genes between the individuals in the reference site and those in the outfall (Figure 53). All results obtained indicate that the induction factors observed more heavily favor an adaptive response of organisms to the disinfection of treated wastewater by PFA rather than an acute toxicity response.

Key points

As regards PFA disinfection performance:

- The disinfection performance of limited doses of PFA (1.2 ppm of PFA and 18 min of contact time) applied to treated wastewater has been confirmed and quantified.
- The average removal for *E. coli* is 3.4 u.log.
- The average removal for *E. faecalis* is 2.2 u.log.
- A systematically lower removal of one log unit for *E. faecalis* compared to *E. coli* has been observed.
- The disinfection effectiveness is highly dependent on the outlet water quality and, in particular, on the TSS load.

In studying the impact of PFA disinfection on the natural environment:

- The WWTP outlet pipe behaves like an artificial reef and contains an ecosystem different from the surrounding sandy seabed.
- Biodiversity appears to be normal given the environmental constraints (swell), and the identified species are those commonly found in the Bay of Biscay.
- A more detailed study on a bio-indicator (common mussel) was conducted from both a physiological (biometric indices) and genetic (comparative response of certain gene stresses) point of view. All results obtained to date suggest that the observed responses are adaptive in nature. Further anatomical-pathological studies would provide additional information on process safety.

Chapter 2

Case of Venice (Italy)

Patrizia Ragazzo and Nicoletta Chiucchini

© Veritas Spa

2.1 INTRODUCTION

2.1.1 The initial context

This project started up in 2005 with the aim of identifying a valid alternative to chlorine in the disinfection of wastewater. Given its tendency to form disinfectant byproducts (DBPs), as early as 2004 Veneto Region Authority had expressed the intention to proceed with extensive chlorine prohibition in wastewater disinfection. The need had therefore emerged to quickly identify a disinfection alternative that, by ensuring bacterial inactivation performance similar to that of chlorine, did not involve byproduct formation. In reviewing the alternatives

doi: 10.2166/9781789062106_0124

available at the time, peracetic acid (PAA) and UV irradiation constituted the main solutions capable of being adopted. However, since the effectiveness of UV treatment is highly influenced by the quality characteristics of effluent and moreover implies high operating and capital costs while PAA entails other noteworthy disadvantages (qualitative or economic), we did not consider either of them to be adequate solutions.

Hence, we laid out two main lines of research aimed at identifying alternatives, one regarding the circumstances of potentially maintaining chlorine use and the other focused on studying the applicability and reliability of a new chemical (performic acid, PFA), produced by the Finnish company Kemira, proposed as a disinfectant in wastewater.

The suitability of PFA in wastewater disinfection was preliminarily assessed at our laboratory by carrying out batch tests on actual matrices consisting of secondary wastewater effluent, and by verifying the effectiveness of PFA on fecal coliforms *Escherichia coli (E. coli)* and enterococci at doses and contact times of 1–5 mg/L and 10–60 min, respectively. Based on the encouraging results shown for PFA effectiveness versus those of more traditional disinfectants, such as PAA and chlorine (Ragazzo *et al.*, 2007), we initiated the first full-scale experiments, the first of which was carried out in winter 2005 and summer 2006 at the *Caorle* municipal wastewater treatment plant (120,000 population equivalent).

These studies were conducted by using both the first production system prototype (Figure 54a and b) and the more heavily automated pilot *Hyproform* system (Figure 54c and d). As part of these investigations, the effectiveness and potential impacts were tested on a number of key effluent quality parameters (e.g., pH, total organic carbon, other chemical and physical parameters) as well as on the indirect ecotoxicity introduced into the environment.

Further study phases examined other WWTPs (Table 23) and years (2011–2018), with the aim of consolidating the knowledge of PFA with respect to disinfection efficacy and oxidative power on organic matter, in addition to direct and indirect toxicity induced by the doses applied, for potential products generated from its decomposition.

Throughout the experimental campaigns on PFA, specific studies were conducted in parallel within the same WWTPs for comparison with other reference wastewater disinfectants, such as chlorine and PAA.

The full-scale studies demonstrated that PFA is an effective and reliable disinfectant for wastewater treatment applications (Ragazzo *et al.*, 2013), and its effectiveness is comparable and greater than that of chlorine hypochlorite (HYP) and PAA (Ragazzo *et al.*, 2020).

This chapter will report some of the main results obtained, in compliance with the confidentiality agreements set forth by the scientific journals that have published our results. A certain amount of the data has been reprocessed in order to focus on specific aspects of PFA disinfection, while other data are presented here for the first time.

2.1.2 Disinfection methods in VERITAS

Ours is a public multi-utility managing 36 municipalities covering the integrated water cycle, servicing a total of some 800,000 inhabitants across a territory visited every year by millions of tourists. Drinking water is produced from both ground (81%) and surface (19%) sources. Wastewater (95 Mm^3/years) is treated in 38 treatment plants (WWTPs) for a total capacity exceeding 1.2 million population equivalent (p.e.). Eleven WWTPs have been designed to serve between 10,000 and 400,000 p.e., six of which encompass bathing locations along the Adriatic coast: Jesolo, Caorle, Eraclea, Cavallino, Lido di Venezia, Chioggia (capacity range: 32,000–160,000 p.e.). The sensitivity of this territory due to the Venice Lagoon and coastal bathing zones requires special attention to wastewater management. For this reason, stringent regulations on both discharged loads and WWTP treatment requirements were drawn up beginning in the early 1970s (D.P.R. 962/1973). The ban on chlorine use in wastewater disinfection is part of these restrictions. The first prohibition dates back to the early 2000s (D.M. 1999) and referred to effluents directly or indirectly discharging into the Venice Lagoon. This interdiction was then extended to the entire region 14 years later, in 2013 (Deliberation no. 107/2009, Veneto Region, 2009). This context prompted the conversion of disinfection treatments in the region toward alternatives, at the time considered to be less impactful. In the case of Veritas, the initial conversions were to the UV technology at the Fusina (400,000 p.e.) and Campalto (130,000 p.e.) WWTPs, as part of an integrated project to adapt infrastructure to safeguard the Venice Lagoon. By extension later on, other WWTP disinfection treatments were upgraded, by adopting the only chemical alternative to chlorine available during those years (2000–2010). To date, 13 installations offer active disinfection: three use UV technologies (130,000–400,000 p.e.), eight disinfect with PAA (2500–105,000 p.e.), and two disinfect with PFA (32000–160,000 p.e.).

2.2 MATERIALS AND METHODS
2.2.1 Studies and the WWTPs

All studies on PFA, from 2005 to 2018, were carried out with both laboratory tests and full-scale experiments. For this purpose, three coastal WWTPs – i.e., Jesolo (160,000 p.e.), Caorle (120,000 p.e.) and Eraclea (32,000 p.e.) – and the remote San Dona'di Piave plant (45,000 p.e.) were used. All plants provided a conventional sequence of treatment (primary clarification, active sludge oxidation-nitrification and denitrification, secondary settling) for municipal wastewater, collected primarily by combined (60%) sewage systems, with insignificant industrial discharges. The effluents were only disinfected during summer in order to respect the national *E. coli* regulation limit of 5000 CFU/100 mL.

The effectiveness of the disinfectants was tested against fecal coliform (FC), *E. coli* (EC) and the more resistant fecal enterococci (FE). Table 23 provides an overview of all the main investigations undertaken.

All the plants were equipped with traditional chicane disinfection channels; moreover, the doses set to guarantee the compliance limit are flow-paced and maintained to be stable through a standardized management/control system.

Performic acid disinfection solution was produced on-site, just before dose application, with the two systems provided by Kemira Oyj: Hyproform (2005/2006) and Desinfix (starting in 2011). Besides the technical and safety aspects, which were greatly improved in the final unit, the major difference among the systems lies in the compositions of the equilibrium solutions produced, with higher oxidant concentration ranges in the Desinfix (PFA 12–15% w/w and H_2O_2 18–20% w/w), with respect to Hyproform (herein denoted PFA-HP) (PFA 8–10% w/w and H_2O_2 11–13% w/w). A detailed technical description of the systems has been reported in Ragazzo *et al.* (2013).

For the other two disinfectants used here as references, commercial solutions with an active substance nominal titer of 12% w/w for hypochlorite and 15% w/w for PAA were applied.

2.2.2 Sampling methods

Laboratory tests focused on bacterial inactivation over time at different disinfectant doses. For this purpose, samples collected at the WWTP outlet were taken to the laboratory for processing through batch trials at various disinfectant concentrations; upon each test, the samples were quenched for oxidant residual and analyzed for bacterial concentrations according to the method outlined in Table 24. Other batch trials were performed in order to correlate the carbon-based disinfectants PFA and PAA with the organic carbon increase, especially its biodegradable component. Deionized water solutions corresponding to disinfected water, ranging from 0.5 to 10 mg/L, were tested and then total organic carbon (TOC), and chemical and biochemical oxygen demand (COD, BOD_5) were analyzed according to the methods shown in Table 24.

During the full-scale studies, the physicochemical parameters (pH, total suspended solids (TSS), COD, nitrite, ammonia, etc.) at the disinfection inlet were evaluated, along with disinfectant effectiveness by means of FC, EC and FE fecal indicators. In some cases, the EC and FE concentrations at various retention times along the disinfection channel were also determined.

The PFA qualitative impact on organic compounds was assessed by monitoring TOC and formate (FA) concentration, both before and after PFA application. Ecotoxicity impacts were assessed using the Vibrio fishery and Daphnia Magna tests.

All analyses were conducted in the laboratory on composite wastewater samples, obtained by collecting in the same bottles equal aliquots of effluents at different time intervals (every 2–3 hours) in relation to the flow rate; sodium thiosulfate had been previously added to the bottles so as to reduce oxidant residuals.

The disinfectant solution was periodically controlled for the concentration of active substances by titration with cerium ammonium sulfate and sodium thiosulfate for PFA-PAA (Greenspan & MacKellar, 1948) and by iodometric titration with sodium thiosulfate for chlorine (UN EN 901:2007). The oxidant residuals at the disinfection outlet were controlled on a daily basis by means of the DPD colorimetric method, according to APHA 4500-Cl G, with and without the prior addition of catalase enzyme to account for the H_2O_2 contribution in the case of peracids. Table 24 summarizes the methods employed at both the full and laboratory scales.

2.2.3 Data processing

The level of effectiveness was analyzed using the non-parametric Kruskal–Wallis test. The ANCOVA test was applied to evaluate the effects of disinfectants, adjusted for doses and contact time. The t-Student and non-parametric Wilcoxon tests were used for comparisons of quality pair data, respectively for normal and non-normal distributions (based on the Shapiro–Wilks test). Associations among variables were assessed by the Spearman rank test. Bacterial inactivation results were modeled using Hom and S-Model (Luukkonen *et al.*, 2015); kinetics parameter values were obtained by means of non-linear regression run with the Microsoft Excel Solver-function GRG nonlinear.

2.3 RESULTS
2.3.1 Experimental set-up

The first full-scale results on PFA effectiveness were obtained in 2005–2006 by using the prototype and pilot production system (Hyproform, denoted here as PFA-HP). From 2011 onwards, all PFA studies were carried out using increasingly more advanced production systems, culminating in what became the definitive Desinfix production system (denoted here as PFA). Moreover, in those plants where HYP and PAA were routinely used, specific management and monitoring criteria were implemented, in addition to adapting frequency and control type, in order to obtain effectiveness data useful in comparisons. Table 25 lists the installations and operational conditions of the full-scale experiments for all disinfectants presented and discussed in this section. All plants serve seaside locations and are characterized by high seasonal and daily load variations; retention time in the disinfection channel, which is consistently adequate at the Caorle and Jesolo WTTPs, was more critical at Eraclea, where values shorter than 10 min were often recorded. Disinfectant doses, previously individuated to meet the 5000 CFU/100 mL *E. coli* target at the outlet, were generally maintained.

2.3.2 Effectiveness

Table 26 provides the median and range of variation of the main qualitative characteristics of the secondary effluents entering the disinfection reactor for each disinfectant applied (Kruskal–Wallis test). Since these effluents stem from the biological oxidation process, the main physicochemical parameters capable of impacting with disinfectants are: pH, total suspended solids, nitrous and ammonium nitrogen, organic compounds (COD, BOD_5), and the bacterial concentration itself.

Qualitative parameter variations were similar to one another except for ammonia, which was high particularly in the case of disinfection with HYP and PFA (nitrite was always below the detection limit of the analytical method: 0.02 mg/L). Ammonia is reported not to interfere with PAA and PFA; with chlorine, it was determined that the disinfection mainly occurred by monochloramine (Ragazzo *et al.*, 2020). Total suspended solids and COD were always below the values reported as interfering with disinfectants; for pH, only PFA could have interfered with the neutral values (Ragazzo *et al.*, 2020). Bacterial concentrations entering the disinfection channel were comparable ($p > 0.05$, Kruskal–Wallis test), with the exception of fecal coliforms and *E. coli* under PFA-HP disinfection (where lower values were recorded compared to other disinfectants, $p < 0.001$); for PFA-HP, the Spearman test found weak correlations with the corresponding bacterial reduction (R~0.52–0.60 for fecal coliforms and *E. coli* respectively, $p < 0.01$).

Weak correlations were also found for PAA and PFA, respectively, between the enterococci and *E. coli* detected at the inlet and corresponding values at the outlet. Only in the case of PFA-HP was a correlation found, albeit weak, between fecal coliforms measured at the outlet and TSS values at the inlet (R~0.52, $p < 0.01$). The results obtained at full scale from 2005 to 2011 are given in Table 27; data are provided in terms of statistical comparison among the bacterial inactivation achieved by the various disinfectants, each one under the respective operating conditions of dose and contact time (via the Kruskal–Wallis test).

PFA appears in both forms: the oldest disinfection solution with a PFA of around 9% w/w (PFA-HP) and the new solution (PFA) containing a PFA of approximately 14% w/w. At contact times between 10 and 20 min, at doses that were one-half and one-third of those of PAA and HYP respectively, PFA achieved the highest inactivation of all fecal indicators, while PFA-HP and PAA the lowest against *E. coli* and enterococci, respectively. With less constraining contact times however, the PFA-HP achieved *E. coli* reductions comparable to those with HYP and PAA, as well as an enterococci inactivation similar to that of HYP; put otherwise, it performs better than or comparable with the other chemical disinfectants.

Under very limited retention times, PFA achieved *E. coli* and enterococci inactivation consistently greater than that of PAA over all time intervals, with enterococci reductions being comparable to those reached by chlorine at 30-plus minutes. Figure 55 provides the exceedance probability plot of disinfectants from 10 to 20 minutes.

The ANCOVA test was applied to these same results to compare the disinfectant performance in taking into account the contribution of doses applied, thus evaluating the actual effectiveness of the disinfectants. The order of effectiveness indicated by this test is summarized in Table 28.

Bacterial inactivation was fitted with the Homs and S bacterial models to compare PFA and PAA; the range of parameters estimated is reported in Table 29. Fecal indicator inactivation has been well described by both models (R^2 always above 0.91) at doses between 0.5 and 1.3 mg/L for PFA (manly at the full scale) and 1–3 mg/L for PAA (mainly at the laboratory scale). Hom's model shows the limited importance of disinfectant concentrations, except for PAA, against enterococci ('n' values greater than the others), as well as the importance of retention time, which increases with bacterial resistance ('m' values greater for Enterococci than *E. coli*) and the disinfectant strength decrease ('m' values less for PFA than PAA). These findings are in agreement with results reported by Luukkonen *et al.* (2015). The K values for both models further confirmed the higher sensitivity of *E. coli* than enterococci to disinfection (values for *E. coli* greater than those for enterococci).

2.3.3 Reuse goal

Results obtained during experiments at the full scale (2005–2011) were analyzed to verify the extent to which the disinfection systems, set for compliance at the current target (5000 *E. coli* CFU/100 mL), were able to meet more ambitious goals. For the four disinfection systems tested, Table 30 reports the percentage of values at the disinfection outlet capable of respecting various limits: from the targets to be respected (the highest value, representing the general limit that effluents are to respect for discharge into surface waters) to the other two more restrictive targets, used as conservative levels to guarantee compliance. In the case of *E. coli*, 1000 CFU/100 mL offers a guideline value by the WHO for agricultural reuse in countries without drought issues; 10 CFU/100 mL must be respected in Italy for this same kind of reuse (D.M. 185/2003).

The ability to comply with the various limits can be summarized in the following order: HYP > PFA-HP > PAA with fecal coliforms; PFA > HYP~PFA-HP > PAA against *E. coli* and enterococci. Moreover, the PFA constantly met the 1000 *E. coli* WHO guideline value and yielded the best performance for the other targets as well.

The ANCOVA model applied to the results discussed above, including doses and contact times (covariates), provides a forecast of disinfectant effectiveness adjusted for both covariates; Table 31 shows how the various systems can guarantee a low level of microbiological risk in the effluent. Among all the bacterial indicators,

PFA is supposed to be the most effective in guaranteeing a low level of microbiological risk, especially with enterococci; the values guaranteed at the discharge outlet by PAA and PFA are respectively 1.5 and 0.2 times those of chlorine. In other words, the microbiological values in the effluent guaranteed by PAA are 7–9 times higher than those guaranteed by PFA (for Fecal coli to *E. coli*, respectively).

2.3.4 Compliance over time

Subsequent to these studies, PFA disinfection technology was adopted in all the WWTPs where the experiments were conducted (Caorle, Eraclea, San Donà e Jesolo). For the several contact time intervals of WWTPs over the six-year period (2013–2018), Table 32 reports the number of controls performed, the corresponding doses applied, and the percentage of results able to meet 1000 CFU/100 mL and the two additional, more ambitious *E. coli* targets. The physicochemical characteristics of the effluent were those typical of the WWTPs: some key parameters are indicated in terms of average, standard deviation and maximum value (in brackets) in Table 33.

The *E. coli* goal (5000 CFU/100 mL) was respected at a high level of confidence (1000 CFU/100 mL) in 99% of controls (1000 CFU/100 mL). When taking into consideration all WWTPs, the PFA doses ranged from a minimum of 0.4 to a maximum of 1.1 mg/L, and retention times in the disinfection channel were typical of each installation, that is relatively high for Caorle and San Donà (52 and 27 min on average, respectively) and quite low at Jesolo and Eraclea (20 and 10 min, respectively). The few values that exceeded 1000 *E. coli* CFU/100 mL were recorded at the Eraclea WWTP during high season. The *E. coli* limit established for Italian agricultural reuse (10 CFU/100 mL) was met in less than 20% of cases across all installations except for San Donà, where this limit was met in 58% of cases.

In the two-year period 2016–2017, PFA residuals at the disinfection outlet were measured three times a day at each installation to monitor the decomposition of oxidant residuals. In 97% of the measurements, the PFA dose was below 0.8 mg/L (total average: 0.6 mg/L), and retention times were shorter than 20 and 40 min in 53 and 74% of the cases, respectively. The average retention times in the disinfection reactors were ranked as follows: Caorle > S. Donà > Jesolo > Eraclea, where the extremes were respectively 52 and 10 min.

The trend in PFA residuals at the disinfection outlet was significantly correlated with retention times in the disinfection channels (R Spearman 0.66 $p < 0.01$), regardless of the matrix variation and uncertainty due to in-field measurements and low concentrations. Unlike PFA, H_2O_2 was consumed rapidly and, regardless of contact time, with residuals ranging from 14 to 10% of doses applied at Caorle and Eraclea, respectively.

Figure 56 shows the average residual concentrations of PFA at the four WWTP disinfection outlets vs. their average contact times.

2.3.5 Quality impacts

Several works in the literature have referred to the low propensity of PFA and PAA to form disinfection byproducts. In reference to other authors, Luukkonen and Pehkonen (2017) postulated, for example, that one of the main advantages of PAA over free chlorine or ozone is the reduced probability of forming DPBs and moreover that PFA, though less widely studied in terms of byproduct formation, does form DBPs similar to or even slightly lower than PAA.

Similarly, several pieces of evidence point to the impacts of PAA on the organic substance of the effluent. Both COD and TOC increase as a result of peracid dosing, and it is possible to calculate the theoretical increase based on the peracid equilibrium composition (Luukkonen *et al.*, 2014).

The impact of PFA on the quality characteristics of secondary effluent was examined several times during our experiments, between 2005 and 2017. Initial studies examined the variation, due to the doses applied, in key quality parameters such as TOC, BOD5 and pH (Ragazzo *et al.*, 2013). Subsequent research on the impacts of PFA have focused on the potential cytotoxic, genotoxic and mutagenic effects of PFA (Ragazzo *et al.*, 2017), as well as on certain implications (linked to its use), such as the potential direct toxicity on the environment and the ability to oxidize some of the categories of organic compounds more recalcitrant to oxidation (Ragazzo *et al.*, 2020).

Given the greater ability of PFA, in comparison with PAA, to inactivate the most resistant fecal indicators (enterococci), the ability of PFA to oxidize organic substances has been constantly assessed, either directly on generic organic substances or through indirect testing (Ragazzo *et al.*, 2013, 2020).

No pH variation or organic matter oxidation was detected at doses applied at the full-scale (Ragazzo *et al.*, 2007, 2013). This finding was subsequently confirmed by additional literature reporting a weak ability for PFA to oxidize pharmaceutical compounds, endocrine disruptors and bisphenol-A (Gagnon *et al.*, 2008; Luukkonen *et al.*, 2015).

For each interval of contact time in the disinfection reactor, Tables 34 and 35 compare TOC and formate by using the Student's and Wilcoxon tests, respectively.

The comparison is drawn between values calculated at the dosage point T_0 (obtained by adding the stoichiometric values derived from the applied PFA dosage to the measured values at the inlet) and those measured at the disinfection reactor outlet.

All values measured at the discharge outlet corresponded to the expected theoretical values ($p > 0.05$), with the exception of two time intervals for TOC, during which the values increased from the disinfection reactor inlet to outlet; this variation however was less than or equal to the uncertainty of the analytical method (5–10%). These results suggest a poor ability of PFA to oxidize the organic substance, even the simplest, co-generated and coexistent with the disinfectant itself, in addition to a potentially low tendency to form byproducts.

Specific laboratory tests carried out to correlate the carbon-based disinfectants PFA-HP and PAA with the biodegradability of corresponding residual molecules showed a poor biodegradability in terms of the BOD5 of formate (co-present and intermediate decomposition product of PFA) with respect to acetate (co-present and intermediate product of decomposition of PAA). Table 36 correlates PFA-HP and PAA doses (applied or simulated) with the several corresponding components of measured organic carbon. In the case of the PFA solutions being used today, these components are reduced to 0.7, 1.0 and 0.3 mg/L for TOC, COD and BOD5, respectively.

The toxic effect of PFA was measured in relation to the potential byproducts formed by its reaction with organic and inorganic substances present in the matrix. Toxicity was evaluated both in terms of eco-toxicity induced on Vibrio Fishery and Daphnia Magna (at the doses applied at full-scale, averaging around 1 mg/L, variation range: 0.4–2.4 mg/L) and in terms of mutagenic, cytotoxic and genotoxic potential, by measuring short-term effects on bacterial, plant and mammalian target cells (doses ranging from 0.6 to 1.5 mg/L). In no cases were any depression effects actually recorded on Vibrio Fishery light emission $\geq 50\%$ (Ragazzo *et al.*, 2013) or on Daphnia Magna mobility (zero effect consistently reported). Light suppression, which varied from a minimum of 0% to a maximum of 43%, did not differ from the typical effluent values in the absence of disinfection treatment ($p > 0.05$).

The mutagenic and genotoxic effects were always negative in all *in-vitro* tests, performed on both disinfected and non-disinfected effluents. In tests with *Allium cepa* however, in some samples of unconcentrated wastewater, treatment with PFA did induce a slight increase in the frequencies of the micronucleus in root cells, uncorrelated with the disinfectant doses (Ragazzo *et al.*, 2017).

2.3.6 Protection outcomes and economic evaluations

Effectiveness levels obtained for PFA and PAA in the studies performed from 2011 to 2017, when examined as a whole, allow some conclusions to be drawn on doses and costs required to guarantee, at least in theory, a suitable level of protection from microbiological risks in wastewater disinfection. For both peracids, the doses were set at the minimum value in order to guarantee stable compliance of 5000 CFU/100 mL *E. coli* at the disinfection outlet, which corresponds to respecting the microbiological targets at 20% of the guideline or limit values (Ragazzo *et al.*, 2020).

Table 37 summarizes the percentage of cases and corresponding operating conditions in which the two peracids did respect the conservative microbiological targets.

The 1000 CFU/100 mL *E. coli* and 400 CFU/100 mL enterococci targets were achieved by PAA-PFA in 92–99 and 48–88% of cases, respectively, at applied doses 2–3 times higher for PAA than PFA. In other words, PFA guaranteed higher compliance against *E. coli* despite being much more often under less favorable contact time conditions; given the rather similar operating conditions against enterococci, PFA provided a level of protection nearly double that of PAA.

Costs vary depending on different factors (allocation, type of WWTP and wastewater characteristics, microbial target, etc.); hence, definitive evaluations cannot be given. Nevertheless, some conclusions can be drawn based on the stability of results reported and discussed by Ragazzo *et al.* (2020). Even though the commercial costs of the PFA active substance are on average roughly 1.5 times those of PAA (€10.8/kg and €7.6/kg, respectively – values extracted from internal and literature data – Luukkonen *et al.*, 2015; Maffettone *et al.*, 2018), in all trials the reagent disinfection costs were higher for PAA compared to the on-site generated PFA, that is €0.018/m^3 and €0.010/m^3 on average, respectively.

Key points

- PFA technology appears to be a valid alternative to chlorine and moreover offers some decisive advantages over peracetic acid disinfection.
- Its undisputed efficacy against enterococci under all operating conditions tested is the main reason why we chose PFA as the disinfection technology for plants with effluent more directly connected to the Adriatic coast and bathing zones protection issues. We therefore regard this technology as a solution to be adopted more extensively in the near future.

General conclusion

Among the chemical solutions for disinfecting wastewater that have emerged over the last few decades, performic acid is an adapted solution for wastewater treatment plant outfalls. This book has consolidated the present state of knowledge and moreover provided evidence of its suitability for wastewater disinfection. The purpose here has been threefold: characterize the performance of chemical disinfection by performic acid, define the most suitable implementation conditions (treatment rate, in particular), and ensure environmental protections when implementing this additional treatment step.

To meet these objectives, the Greater Paris Sanitation Authority (SIAAP) and its scientific and industrial partners undertook a two-year research effort on two different scales. On the one hand, tests were carried out at the laboratory scale to define the performance of this disinfection method and, in particular, to identify the links between its effectiveness and implementation conditions. This laboratory scale provided technical and scientific information on the harmlessness of such treatment for the environment, using a large panel of bioassays. On the other hand, trials were conducted at an industrial scale. Disinfection facilities, allowing for the injection of performic acid, were set up at the Seine Valenton wastewater treatment plant (SIAAP, Valenton) in order to assess the performance of this technology in an industrial application while also confirming the lack of impact from the disinfected water on the river.

The effectiveness of performic acid in removing fecal bacteria has been demonstrated. The application of a dose of 1 ppm, corresponding to a $C \times t$ ranging from 10 to 30 ppm.min in the Seine Valenton configuration, made it possible to maintain, under all circumstances, fecal bacteria concentrations below the bathing quality thresholds (Directive 2006/7/EC). The introduction of a wide array of biological tools has indeed proven the lack of harmful effects from performic acid within the aquatic environment regarding endocrine disruption (thyroid, estrogenic, androgenic) and so-called general toxicity (effect on the growth of single-cell organisms). The fact that this disinfection method leads to irreversible damage to bacteria, along with its ability to remove some of the other pathogens (sulfate-reducing spores), yields an even broader range of uses.

We are therefore hoping that this book will assist stakeholders and operators in making choices and managing their wastewater flows in light of changes in social expectations and, more broadly, in offering solutions to mitigate water-related challenges through guidance on sustainable use or reuse.

References

General introduction

Rocher V. and Azimi S. (2016). Microbial Quality of Waters in the Paris Area: From Wastewaters to Surface Waters. Johanet Publisher, Paris. 94 p. ISBN: 979-10-91089-29-6.

Rocher V. and Azimi S. (2017). Improvement of the River Seine Quality in Connection with the Sanitation Changes: From 1975 to 2015. Johanet Publisher, Paris. 76 p. ISBN: 979-10-91089-31-9.

Section 1 – Chapter 1

Chhetri R. K., Thornberg D., Berner J., Gramstad R., Öjstedt U., Sharma A. K. and Andersen H. R. (2014). Chemical disinfection of combined sewer overflow waters using performic acid or peracetic acids. *Science of the Total Environment*, **490**, 1065–1072. https://doi.org/10.1016/j.scitotenv.2014.05.079.

Chhetri R. K., Flagstad R., Munch E. S., Hørning C., Berner J., Kolte-Olsen A., Thornberg D. and Andersen H. R. (2015). Full scale evaluation of combined sewer overflows disinfection using performic acid in a sea-outfall pipe. *Chemical Engineering Journal*, **270**, 133–139. https://doi.org/10.1016/j.cej.2015.01.136.

Chhetri R. K., Klupsch E., Andersen H. R. and Jensen P. E. (2018). Treatment of Arctic wastewater by chemical oagulation, UV and peracetic acid disinfection. *Environmental Science and Pollution Research International*, **25**, 32851–32859. https://doi.org/10.1007/s11356-017-8585-5.

Directive 2006/7/EC of the European Parliament and of the Council of 15 February 2006 concerning the management of bathing water quality and repealing Directive 76/160/EEC. Official Journal of the European Union, L 64/37, 15 pages.

Gehr R., Chen D. and Moreau M. (2009). Performic acid (PFA): tests on an advanced primary effluent show promising disinfection performance. *Water Science and Technology*, **59**, 89–96. https://doi.org/10.2166/wst.2009.761.

Goffin A., Guérin S., Rocher V. and Varrault G. (2018). Towards a better control of the wastewater treatment process: excitation-emission matrix fluorescence spectroscopy of dissolved organic matter as a predictive tool of soluble BOD5 in influents of six

Parisian wastewater treatment plants. *Environmental Science and Pollution Research*, **25**(10), 8765–8776. https://doi.org/10.1007/s11356-018-1205-1.

Karpova T., Pekonen P., Gramstad R., Öjstedt U., Laborda S., Heinonen-Tanski H., Chávez A. and Jiménez B. (2013). Performic acid for advanced wastewater disinfection. *Water Science and Technology*, **68**, 2090–2096. https://doi.org/10.2166/wst.2013.468.

Luukkonen T. and Pehkonen S. O. (2017). Peracids in water treatment: a critical review. *Critical Reviews in Environmental Science and Technology*, **47**, 1–39. https://doi.org/10.1080/10643389.2016.1272343.

Luukkonen T., Heyninck T., Rämö J. and Lassi U. (2015). Comparison of organic peracids in wastewater treatment: disinfection, oxidation and corrosion. *Water Research*, **85**, 275–285. https://doi.org/10.1016/j.watres.2015.08.037.

McFadden M., Loconsole J., Schockling A. J., Nerenberg R. and Pavissich J. P. (2017). Comparing peracetic acid and hypochlorite for disinfection of combined sewer overflows: effects of suspended-solids and pH. *Science of the Total Environment*, **599–600**, 533–539. https://doi.org/10.1016/j.scitotenv.2017.04.179.

Mora M., Veijalainen A.-M. and Heinonen-Tanski H. (2018). Performic acid controls better Clostridium tyrobutyricum related bacteria than peracetic acid. *Sustainability*, **10**, 1–8.

Passerat J., Ouattara N. K., Mouchel J.-M., Vincent R. and Servais P. (2011). Impact of an intense combined sewer overflow event on the microbiological water quality of the Seine River. *Water Research*, **45**, 893–903. https://dx.doi.org/10.1016/j.watres.2010.09.024.

Ragazzo P., Chiucchini N., Piccolo V. and Ostoich M. (2013). A new disinfection system for wastewater treatment: performic acid full-scale trial evaluations. *Water Science and Technology*, **67**, 2476–2487. https://doi.org/10.2166/wst.2013.137.

Ragazzo P., Feretti D., Monarca S., Dominici L., Ceretti E., Viola G., Piccolo V., Chiucchini N. and Villarini M. (2017). Evaluation of cytotoxicity, genotoxicity, and apoptosis of wastewater before and after disinfection with performic acid. *Water Research*, **116**, 44–52. https://doi.org/10.1016/j.watres.2017.03.016.

Rocher V. and Azimi S. (2016). Microbial Quality of Waters in the Paris Area: From Wastewaters to Surface Waters. Johanet Publisher, Paris. 94 pages. ISBN: 979-10-91089-29-6.

Rocher V., Paffoni C., Gonçalves A., Guérin S., Azimi S., Gasperi J., Moilleron R. and Pauss A. (2012). Municipal wastewater treatment by biofiltration: comparisons of various treatment layouts. Part 1: assessment of carbon and nitrogen removal. *Water Science and Technology*, **65**, 1705–17--12. https://doi.org/10.2166/wst.2012.105.

Tondera K., Klaer K., Koch C., Hamza I. A. and Pinnekamp J. (2016). Reducing pathogens in combined sewer overflows using performic acid. *International Journal of Hygiene and Environmental Health*, **219**, 700–708. https://doi.org/10.1016/j.ijheh.2016.04.009.

Section 1 – Chapter 2

Antonelli M., Turolla A., Mezzanotte V. and Nurizzo C. (2013). Peracetic acid for secondary effluent disinfection: a comprehensive performance assessment. *Water Science and Technology*, **68**, 2638–2644. https://doi.org/10.2166/wst.2013.542.

Biancullo F., Moreira N. F. F., Ribeiro A. R., Manaia C. M., Faria J. L., Nunes O. C., Castro-Silva S. M. and Silva A. M. T. (2019). Heterogeneous photocatalysis using UVA-LEDs for the removal of antibiotics and antibiotic resistant bacteria from urban wastewater treatment plant effluents. *Chemical Engineering Journal*, **367**, 304–313. https://doi.org/10.1016/j.cej.2019.02.012.

Bohrerova Z. and Linden K. G. (2007). Standardizing photoreactivation: comparison of DNA photorepair rate in Escherichia coli using four different fluorescent lamps. *Water Research*, **41**, 2832–2838. https://doi.org/10.1016/j.watres.2007.03.015.

Bohrerova Z., Rosenblum J. and Linden K. G. (2015). Importance of recovery of *E. coli* in water following ultraviolet light disinfection. *Journal of Environmental Engineering*, **141**, 04014094. https://doi.org/10.1061/(ASCE)EE.1943-7870.0000922.

Fiorentino A., Ferro G., Alferez M. C., Polo-López M. I., Fernández-Ibañez P. and Rizzo L. (2015). Inactivation and regrowth of multidrug resistant bacteria in urban wastewater after disinfection by solar-driven and chlorination processes. *Journal of Photochemistry and Photobiology B: Biology*, **148**, 43–50. https://doi.org/10.1016/j.jphotobiol.2015.03.029.

Giannakis S., Darakas E., Escalas-Cañellas A. and Pulgarin C. (2015). Solar disinfection modeling and post-irradiation response of Escherichia coli in wastewater. *Chemical Engineering Journal*, **281**, 588–598. https://doi.org/10.1016/j.cej.2015.06.077.

Giannakis S., Voumard M., Grandjean D., Magnet A., De Alencastro L. F. and Pulgarin C. (2016). Micropollutant degradation, bacterial inactivation and regrowth risk in wastewater effluents: influence of the secondary (pre)treatment on the effectiveness of Advanced Oxidation Processes. *Water Research*, **102**, 505–515. https://doi.org/10.1016/j.watres.2016.06.066.

Häder D.-P., Williamson C. E., Wängberg S.-Å., Rautio M., Rose K. C., Gao K., Helbling E. W., Sinha R. P. and Worrest R. (2015). Effects of UV radiation on aquatic ecosystems and interactions with other environmental factors. *Photochemical and Photobiological Science*, **14**, 108–126. https://doi.org/10.1039/C4PP90035A.

Li D., Zeng S., Gu A. Z., He M. and Shi H. (2013). Inactivation, reactivation and regrowth of indigenous bacteria in reclaimed water after chlorine disinfection of a municipal wastewater treatment plant. *Journal of Environmental Science*, **25**, 1319–1325. https://doi.org/10.1016/S1001-0742(12)60176-4.

Malvestiti J. A. and Dantas R. F. (2018). Disinfection of secondary effluents by O-3, O-3/H2O2 and UV/H2O2: influence of carbonate, nitrate, industrial contaminants and regrowth. *Journal of Environmental Chemical Engineering*, **6**, 560–567. https://doi.org/10.1016/j.jece.2017.12.058.

Maraccini P. A., Mattioli M. C. M., Sassoubre L. M., Cao Y., Griffith J. F., Ervin J. S., Van De Werfhorst L. C. and Boehm A. B. (2016). Solar inactivation of Enterococci and *Escherichia coli* in natural waters: effects of water absorbance and depth. *Environmental Science and Technology*, **50**, 5068–5076. https://doi.org/10.1021/acs.est.6b00505.

Mecha A. C., Onyango M. S., Ochieng A. and Momba M. N. B. (2017). Evaluation of synergy and bacterial regrowth in photocatalytic ozonation disinfection of municipal wastewater. *Science of the Total Environment*, **601–602**, 626–635. https://doi.org/10.1016/j.scitotenv.2017.05.204.

Muela A., García-Bringas J. M., Arana I. and Barcina I. (2000). The effect of simulated solar radiation on *Escherichia coli*: the relative roles of UV-B, UV-A, and photosynthetically active radiation. *Microbial Ecology*, **39**, 65–71. https://doi.org/10.1007/s002489900181.

Oguma K., Katayama H. and Ohgaki S. (2002). Photoreactivation of *Escherichia coli* after low- or medium-pressure UV disinfection determined by an endonuclease sensitive site assay. *Applied and Environmental Microbiology*, **68**, 6029–6035. https://doi. org/10.1128/AEM.68.12.6029-6035.2002.

Rocher V. and Azimi S. (2016). Microbial Quality of Waters in the Paris Area: From Wastewaters to Surface Waters. Johanet Publisher, Paris. 94 pages. ISBN: 979-10-91089-29-6.

Shang C., Cheung L. M., Ho C.-M. and Zeng M. (2009). Repression of photoreactivation and dark repair of coliform bacteria by TiO2-modified UV-C disinfection. *Applied Catalysis B: Environmental*, **89**, 536–542. https://doi.org/10.1016/j.apcatb. 2009.01.020.

Voet D. and Voet J. G. (1995). Biochemistry. J. Wiley & Sons, New York, pp. 445–466.

Whitton R., Fane S., Jarvis P., Tupper M., Raffin M., Coulon F. and Nocker A. (2018). Flow cytometry-based evaluation of the bacterial removal effectiveness of a blackwater reuse treatment plant and the microbiological changes in the associated non-potable distribution network. *Science of the Total Environment*, **645**, 1620–1629. https://doi. org/10.1016/j.scitotenv.2018.07.121.

Zhang C., Brown P. J. B. and Hu Z. (2019a). Higher functionality of bacterial plasmid DNA in water after peracetic acid disinfection compared with chlorination. *Science of the Total Environment*, **685**, 419–427. https://doi.org/10.1016/j.scitotenv.2019.05.074.

Zhang C., Brown P. J. B., Miles R. J., White T. A., Grant D. G., Stalla D. and Hu Z. (2019b). Inhibition of regrowth of planktonic and biofilm bacteria after peracetic acid disinfection. *Water Research*, **149**, 640–649. https://doi.org/10.1016/j.watres.2018. 10.062.

Zhou X., Li Z., Lan J., Yan Y. and Zhu N. (2017). Kinetics of inactivation and photoreactivation of Escherichia coli using ultrasound-enhanced UV-C light-emitting diodes disinfection. *Ultrasonics Sonochemistry*, **35**, 471–477. https://doi.org/10. 1016/j.ultsonch.2016.10.028.

Section 1 – Chapter 3

Alberts J. J. and Takács M. (2004). Comparison of the natural fluorescence distribution among size fractions of terrestrial fulvic and humic acids and aquatic natural organic matter. *Organic Geochemistry*. **35**, 1141–1149. https://doi.org/10.1016/j. orggeochem.2004.06.010.

Barreto J. C., Smith G. C., Strobel N. H. P., McQuillin P. A. and Miller T. A. (1994). Terephthalic acid: a dosimeter for the detection of hydroxyl radicals in vitro. *Life Science*, **56**(4), 89–96.

Determann S., Lobbes J. M., Reuter R. and Rullkötter J. (1998). Ultraviolet fluorescence excitation and emission spectroscopy of marine algae and bacteria. *Marine Chemistry*, **62**(1), 137–156.

Domínguez Henao L., Cascio M., Turolla A. and Antonelli M. (2018). Effect of suspended solids on peracetic acid decay and bacterial inactivation kinetics: experimental assessment and definition of predictive models. *Science of the Total Environment*, **643**, 936–945.

Filippis P. D., Scarsella M. and Verdone N. (2009). Peroxyformic acid formation: a kinetic study. *Industrial & Engineering Chemistry Research*, **48**(3), 1372–1375.

Lakowicz J. R. (2010). Principles of Fluorescence Spectroscopy, 4th edn. Springer Science and Business Media, New York, NY.

Lawaetz A. J. and Stedmon C. A. (2009). Fluorescence intensity calibration using the raman scatter peak of water. *Applied Spectroscopy*, **63**(8), 936–940.

Luukkonen T. and Pehkonen S. O. (2017). Peracids in water treatment: a critical review. *Critical Reviews in Environmental Science and Technology*, **47**(1), 1–39.

Luukkonen T., Heyninck T., Rämö J. and Lassi U. (2015). Comparison of organic peracids in wastewater treatment: disinfection, oxidation and corrosion. *Water Research* **85**, 275–285.

Parlanti E., Wörz K., Geoffroy L. and Lamotte M. (2000). Dissolved organic matter fluorescence spectroscopy as a tool to estimate biological activity in a coastal zone submitted to anthropogenic inputs. *Organic Geochemistry*, **31**(12), 1765–1781.

Pinkernell U., Lüke H.-J. and Karst U. (1997). Selective photometric determination of peroxycarboxylic acids in the presence of hydrogen peroxide. *Analyst*, **122**(6), 567–571.

Santacesaria E., Russo V., Tesser R., Turco R. and Di Serio M. (2017). Kinetics of performic acid synthesis and decomposition. *Industrial & Engineering Chemistry Research*, **56**(45), 12940–12952.

Watras C. J., Hanson P. C., Stacy T. L., Morrison K. M., Mather J., Hu Y.-H. and Milewski P. (2011). A temperature compensation method for CDOM fluorescence sensors in freshwater. *Limnology and Oceanography: Methods*, **9**(7), 296–301.

Wenk J., Von Gunten U. and Canonica S. (2011). Effect of dissolved organic matter on the transformation of contaminants induced by excited triplet states and the hydroxyl radical. *Environmental Science and Technology*, **45**(4), 1334–1340.

Zhang W., Cao B., Wang D., Ma T., Xia H. and Yu D. (2016). Influence of wastewater sludge treatment using combined peroxyacetic acid oxidation and inorganic coagulants re-flocculation on characteristics of extracellular polymeric substances (EPS). *Water Research*, **88**, 728–739.

Section 1 – Chapter 4

Alexandrou L., Meehan B. J. and Jones O. A. H. (2018). Regulated and emerging disinfection by-products in recycled waters. *Science of the Total Environment*, **637–638**, 1607–1616. https://doi.org/10.1016/j.scitotenv.2018.04.391.

Bergé A., Cladière M., Gasperi J., Coursimault A., Tassin B. and Moilleron R. (2012). Meta-analysis of environmental contamination by alkylphenols. *Environmental Science and Pollution Research*, **19**, 3798–3819. https://doi.org/10.1007/s11356-012-1094-7.

Bergé A., Buleté A., Fildier A., Mailler R., Gasperi J., Coquet Y., Nauleau F., Rocher V. and Vulliet E. (2018). Non-target strategies by HRMS to evaluate fluidized micro-grain activated carbon as a tertiary treatment of wastewater. *Chemosphere*, **213**, 587–595. https://doi.org/10.1016/j.chemosphere.2018.09.101.

Boccard J., Veuthey J.-L. and Rudaz S. (2010). Knowledge discovery in metabolomics: an overview of MS data handling. *Journal of Separation Science*, **33**, 290–304. https://doi.org/10.1002/jssc.200900609.

Bond T., Huang J., Templeton M. R. and Graham N. (2011). Occurrence and control of nitrogenous disinfection by-products in drinking water – a review. *Water Research*, **45**, 4341–4354. https://doi.org/10.1016/j.watres.2011.05.034.

Chhetri R. K., Thornberg D., Berner J., Gramstad R., Öjstedt U., Sharma A. K. and Andersen H. R. (2014). Chemical disinfection of combined sewer overflow waters using performic acid or peracetic acids. *Science of the Total Environment*, **490**, 1065–1072. https://doi.org/10.1016/j.scitotenv.2014.05.079.

Dell'Erba A., Falsanisi D., Liberti L., Notarnicola M. and Santoro D. (2007). Disinfection by-products formation during wastewater disinfection with peracetic acid. *Desalination*, **215**, 177–186. https://doi.org/10.1016/j.desal.2006.08.021.

Domínguez Henao L., Turolla A. and Antonelli M. (2018). Disinfection by-products formation and ecotoxicological effects of effluents treated with peracetic acid: a review. *Chemosphere*, **213**, 25–40. https://doi.org/10.1016/j.chemosphere.2018.09.005.

Gerrity D., Pisarenko A. N., Marti E., Trenholm R. A., Gerringer F., Reungoat J. and Dickenson E. (2015). Nitrosamines in pilot-scale and full-scale wastewater treatment plants with ozonation. *Water Research*, **72**, 251–261. https://doi.org/10.1016/j.watres.2014.06.025.

Glover C. M., Verdugo E. M., Trenholm R. A. and Dickenson E. R. V. (2019). N-nitrosomorpholine in potable reuse. *Water Research*, **148**, 306–313. https://doi.org/10.1016/j.watres.2018.10.010.

Guillossou R., Le Roux J., Mailler R., Vulliet E., Morlay C., Nauleau F., Gasperi J. and Rocher V. (2019). Organic micropollutants in a large wastewater treatment plant: what are the benefits of an advanced treatment by activated carbon adsorption in comparison to conventional treatment? *Chemosphere*, **218**, 1050–1060. https://doi.org/10.1016/j.chemosphere.2018.11.182.

Heeb M. B., Criquet J., Zimmermann-Steffens S. G. and von Gunten U. (2014). Oxidative treatment of bromide-containing waters: formation of bromine and its reactions with inorganic and organic compounds – A critical review. *Water Research*, **48**, 15–42. https://doi.org/10.1016/j.watres.2013.08.030.

Held A. M., Halko D. J. and Hurst J. K. (1978). Mechanisms of chlorine oxidation of hydrogen peroxide. *Journal of the American Chemical Society*, **100**, 5732–5740. https://doi.org/10.1021/ja00486a025.

Hogenboom A. C., van Leerdam J. A. and de Voogt P. (2009). Accurate mass screening and identification of emerging contaminants in environmental samples by liquid chromatography–hybrid linear ion trap Orbitrap mass spectrometry. *Journal of Chromatography A*, **1216**, 510–519. https://doi.org/10.1016/j.chroma.2008.08.053.

Ibáñez M., Sancho J. V., Hernández F., McMillan D. and Rao R. (2008). Rapid non-target screening of organic pollutants in water by ultraperformance liquid chromatography coupled to time-of-light mass spectrometry. *Trends in Analytical Chemistry*, **27**, 481–489. https://doi.org/10.1016/j.trac.2008.03.007.

Keefer L. K. and Roller P. P. (1973). N-nitrosation by nitrite ion in neutral and basic medium. *Science*, **181**, 1245–1247.

Kitis M. (2004). Disinfection of wastewater with peracetic acid: a review. *Environment International*, **30**, 47–55. https://doi.org/10.1016/S0160-4120(03)00147-8.

Krasner S. W., Weinberg H. S., Richardson S. D., Pastor S. J., Chinn R., Sclimenti M. J., Onstad G. D. and Thruston A. D. (2006). Occurrence of a new generation of disinfection byproducts. *Environmental Science and Technology*, **40**, 7175–7185. https://doi.org/10.1021/es060353j.

Krauss M., Longrée P., Dorusch F., Ort C. and Hollender J. (2009). Occurrence and removal of N-nitrosamines in wastewater treatment plants. *Water Research*, **43**, 4381–4391. https://doi.org/10.1016/j.watres.2009.06.048.

Krauss M., Singer H. and Hollender J. (2010). LC–high resolution MS in environmental analysis: from target screening to the identification of unknowns. *Analytical and Bioanalytical Chemistry*, **397**, 943–951. https://doi.org/10.1007/s00216-010-3608-9.

Lee J.-H. and Oh J.-E. (2016). A comprehensive survey on the occurrence and fate of nitrosamines in sewage treatment plants and water environment. *Science of the Total Environment*, **556**, 330–337. https://doi.org/10.1016/j.scitotenv.2016.02.090.

Le Roux J., Gallard H. and Croué J.-P. (2011). Chloramination of nitrogenous contaminants (pharmaceuticals and pesticides): NDMA and halogenated DBPs formation. *Water Research*, **45**, 3164–3174. https://doi.org/10.1016/j.watres.2011.03.035.

Mailler R., Gasperi J., Coquet Y., Buleté A., Vulliet E., Deshayes S., Zedek S., Mirande-Bret C., Eudes V., Bressy A., Caupos E., Moilleron R., Chebbo G. and Rocher V. (2016). Removal of a wide range of emerging pollutants from wastewater treatment plant discharges by micro-grain activated carbon in fluidized bed as tertiary treatment at large pilot scale. *Science of the Total Environment*, **542**(Part A), 983–996. https://doi.org/10.1016/j.scitotenv.2015.10.153.

Mailler R., Gasperi J., Vulliet E., Buleté A., Azimi S. and Rocher V. (2017). Evaluation of the footprint of pharmaceutical residues and other emergent pollutants in the wastewaters of the Paris urban area (In French). *L'Eau, L'Industrie, Les Nuisances*, **401**, 90–96

Merel S., Lege S., Yanez Heras J. E. and Zwiener C. (2017). Assessment of N-oxide formation during wastewater ozonation. *Environmental Science and Technology*, **51**, 410–417. https://doi.org/10.1021/acs.est.6b02373.

Mirvish S. S. (1975). Formation of N-nitroso compounds: chemistry, kinetics, and in vivo occurrence. *Toxicology and Applied Pharmacology*, **31**, 325–351. https://doi.org/10.1016/0041-008X(75)90255-0.

Mitch W. A., Sharp J. O., Trussell R. R., Valentine R. L., Alvarez-Cohen L. and Sedlak D. L. (2003). N-nitrosodimethylamine (NDMA) as a drinking water contaminant: a review. *Environmental Engineering Science*, **20**, 389–404. https://doi.org/10.1089/109287503768335896.

Müller A., Schulz W., Ruck W. K. L. and Weber W. H. (2011). A new approach to data evaluation in the non-target screening of organic trace substances in water analysis. *Chemosphere*, **85**, 1211–1219. https://doi.org/10.1016/j.chemosphere.2011.07.009.

Nürenberg G., Kunkel U., Wick A., Falås P., Joss A. and Ternes T. A. (2019). Nontarget analysis: a new tool for the evaluation of wastewater processes. *Water Research*, **163**, 114842. https://doi.org/10.1016/j.watres.2019.07.009.

Park S.-H., Wei S., Mizaikoff B., Taylor A. E., Favero C. and Huang C.-H. (2009). Degradation of amine-based water treatment polymers during chloramination as N-nitrosodimethylamine (NDMA) precursors. *Environmental Science and Technology*, **43**, 1360–1366. https://doi.org/10.1021/es802732z.

Plewa M. J., Wagner E. D., Muellner M. G., Hsu K.-M. and Richardson S. D. (2008). Comparative mammalian cell toxicity of N-DBPs and C-DBPs. In: Disinfection By-products in Drinking Water: Occurrence, Formation, Health Effects, and Control, T. Karanfil, S. W. Krasner, P. Westerhoff and Y. Xie (eds), American Chemical Society, Washington, DC, pp. 36–50. https://doi.org/10.1021/bk-2008-0995.ch003.

Poulin R. X. and Pohnert G. (2018). Simplifying the complex: metabolomics approaches in chemical ecology. *Analytical and Bioanalytical Chemistry*, **411**, 13–19. https://doi.org/10.1007/s00216-018-1470-3.

Ragazzo P., Chiucchini N., Piccolo V. and Ostoich M. (2013). A new disinfection system for wastewater treatment: performic acid full-scale trial evaluations. *Water Science and Technology*, **67**, 2476–2487. https://doi.org/10.2166/wst.2013.137.

Ramadan Z., Jacobs D., Grigorov M. and Kochhar S. (2006). Metabolic profiling using principal component analysis, discriminant partial least squares, and genetic algorithms. *Talanta*, **68**, 1683–1691. https://doi.org/10.1016/j.talanta.2005.08.042.

R Core Team (2019). R: A Language and Environment for Statistical Computing. R Foundation for Statistical Computing, Vienna, Austria.

Richardson S. D., Plewa M. J., Wagner E. D., Schoeny R. and DeMarini D. M. (2007). Occurrence, genotoxicity, and carcinogenicity of regulated and emerging disinfection by-products in drinking water: a review and roadmap for research. *Mutation Research/Reviews in Mutation Research, The Sources and Potential Hazards of Mutagens in Complex Environmental Matrices – Part II*, **636**, 178–242. https://doi.org/10.1016/j.mrrev.2007.09.001.

Ritchie M. E., Phipson B., Wu D., Hu Y., Law C. W., Shi W. and Smyth G. K. (2015). Limma powers differential expression analyses for RNA-sequencing and microarray studies. *Nucleic Acids Research*, **43**, e47.

Schollée J. E., Schymanski E. L. and Hollender J. (2016). Statistical approaches for LC-HRMS data to characterize, prioritize, and identify transformation products from water treatment processes. Assessing Transformation Products of Chemicals by Non-Target and Suspect Screening – Strategies and Workflows Volume 1, ACS Symposium Series, American Chemical Society, Washington, DC, pp. 45–65. https://doi.org/10.1021/bk-2016-1241.ch004.

Schreiber I. M. and Mitch W. A. (2007). Enhanced nitrogenous disinfection byproduct formation near the breakpoint: implications for nitrification control. *Environmental Science and Technology*, **41**, 7039–7046. https://doi.org/10.1021/es070500t.

Schymanski E. L., Singer H. P., Slobodnik J., Ipolyi I. M., Oswald P., Krauss M., Schulze T., Haglund P., Letzel T., Grosse S., Thomaidis N. S., Bletsou A., Zwiener C., Ibáñez M., Portolés T., de Boer R., Reid M. J., Onghena M., Kunkel U., Schulz W., Guillon A., Noyon N., Leroy G., Bados P., Bogialli S., Stipaničev D., Rostkowski P. and Hollender J. (2015). Non-target screening with high-resolution mass spectrometry: critical review using a collaborative trial on water analysis. *Analytical and Bioanalytical Chemistry*, **407**, 6237–6255. https://doi.org/10.1007/s00216-015-8681-7.

Shah A. D. and Mitch W. A. (2012). Halonitroalkanes, halonitriles, haloamides, and N-nitrosamines: a critical review of nitrogenous disinfection byproduct formation pathways. *Environmental Science and Technology*, **46**, 119–131. https://doi.org/10.1021/es203312s.

Shah A. D., Liu Z.-Q., Salhi E., Höfer T. and von Gunten U. (2015). Peracetic acid oxidation of saline waters in the absence and presence of H_2O_2: secondary oxidant and disinfection byproduct formation. *Environmental Science and Technology*, **49**, 1698–1705. https://doi.org/10.1021/es503920n.

Singer H. P., Wössner A. E., McArdell C. S. and Fenner K. (2016). Rapid screening for exposure to "non-target" pharmaceuticals from wastewater effluents by combining HRMS-based suspect screening and exposure modeling. *Environmental Science and Technology*, **50**(13), 6698–6707. https://doi.org/10.1021/acs.est.5b03332.

Wagner E. D. and Plewa M. J. (2017). CHO cell cytotoxicity and genotoxicity analyses of disinfection by-products: an updated review. *Journal of Environmental Sciences*, **58**, 64–76. https://doi.org/10.1016/j.jes.2017.04.021.

West D. M., Wu Q., Donovan A., Shi H., Ma Y., Jiang H. and Wang J. (2016). N-nitrosamine formation by monochloramine, free chlorine, and peracetic acid disinfection with presence of amine precursors in drinking water system. *Chemosphere*, **153**, 521–527. https://doi.org/10.1016/j.chemosphere.2016.03.035.

Wilkinson L. (2011) venneuler: Venn and Euler Diagrams. R package version 1.1-0. https://CRAN.R-project.org/package=venneuler.

Yoon S., Nakada N. and Tanaka H. (2012). A new method for quantifying N-nitrosamines in wastewater samples by gas chromatography – triple quadrupole mass spectrometry. *Talanta*, **97**, 256–261. https://doi.org/10.1016/j.talanta.2012.04.027.

Section 2 – Chapter 1

Chhetri R. K., Flagstad R., Munch E. S., Hørning C., Berner J., Kolte-Olsen A., Thornberg D. and Andersen H. R. (2015). Full scale evaluation of combined sewer overflows disinfection using performic acid in a sea-outfall pipe. *Chemical Engineering Journal*, **270**, 133–139. https://doi.org/10.1016/j.cej.2015.01.136.

Pigot T., de Casamajor M. N., Sanchez F., Gonzalez P. and Paulin T. (2019). Disinfection of the treated effluent of the Biarritz wastewater treatement plant using DesinFix: assessment after 38 months. Internal Report.

Ragazzo P., Chiucchini N., Piccolo V. and Ostoich M. (2013). A new disinfection system for wastewater treatment: performic acid full-scale trial evaluations. *Water Science and Technology*, **67**, 2476–2487. https://doi.org/10.2166/wst.2013.137.

Section 2 – Chapter 2

Angelescu D. E. and Hausot A. (2019). Automating *E. coli* quantification in wastewater and surface waters. *Water Industry Journal*, **December issue** (13), 36–37.

Angelescu D. E. and Saison O. (2020). Innovative Approaches to Study Microbial Impact of Housing Boats. Carrefour de l'Eau conference, Rennes, January 28–31.

Angelescu D. E., Huynh V., Hausot A., Yalkin G., Plet V., Mouchel J. M., Guérin-Rechdaoui S., Azimi S. and Rocher V. (2018a). Autonomous system for rapid field quantification of *Escherichia coli* in surface waters. *Journal of Applied Microbiology*, **126**, 332–343.

Angelescu D. E., Hausot A., Huynh V. and Wong J. (2018b). An in-situ autonomous bacterial pathogen sensor for water quality and environmental monitoring applications. *Proceedings of the Water Environment Federation*, **2018**(9), 4392–4403.

Baker A., Cumberland S. A., Bradley C., Buckley C. and Bridgeman J. (2015). To what extent can portable fluorescence spectroscopy be used in the real-time assessment of microbial water quality? *Science of the Total Environment*, **532**, 14–19.

Baudart J. and Lebaron P. (2010). Rapid detection of *Escherichia coli* in waters using fluorescent in situ hybridization, direct viable counting and solid phase cytometry. *Journal of Applied Microbiology*, **109**, 1253–1264.

Baudart J., Servais P., De Paoli H., Henry A. and Lebaron P. (2009). Rapid enumeration of *Escherichia coli* in marine bathing waters: potential interference of nontarget bacteria. *Journal of Applied Microbiology*, **107**, 2054–2062.

Bergeron P., Oujati H., Cuenca V. C., Mestre J. M. H. and Courtois S. (2011). Rapid monitoring of *Escherichia coli* and *Enterococcus* spp. in bathing water using Reverse Transcription-quantitative PCR. *International Journal of Hygiene and Environmental Health*, **214**, 478–484.

Briciu-Burghina C., Heery B. and Regan F. (2015). Continuous fluorometric method for measuring β-glucuronidase activity: comparative analysis of three fluorogenic substrates. *Analyst*, **140**(17), 5953–5964.

Burnet J.-B., Dinh Q. T., Imbeault S., Servais P., Dorner S. and Prévost M. (2019). Autonomous online measurement of beta-D-glucuronidase activity in surface water: is it suitable for rapid *E. coli* monitoring? *Water Research*, **152**, 241–250.

Cronin T., Loewenthal M., Huynh V., Huynh V., Angelescu D. E. and Hausot A. (2018). Rapid *E. coli* quantification with field portable devices around UK bathing sites. *Water Industry Journal*, 26–27.

Edberg S. C., Rice E. W., Karlin R. J. and Allen M. J. (2000). *Escherichia coli*: the best biological drinking water indicator for public health protection. *Journal of Applied Microbiology*, **88**(Suppl. 1), 106S–116S.

EPA Method 9131 (1986). Total coliform: multiple tube fermentation technique. Available from: https://www.epa.gov/sites/production/files/2015-12/documents/9131.pdf (accessed 16 April 2020).

Heery B., Briciu-Burghina C., Zhang D., Duffy G., Brabazon D., O'Connor N. and Regan F. (2016). ColiSense, today's sample today: a rapid on-site detection of β-d-Glucuronidase activity in surface water as a surrogate for *E. coli*. *Talanta*, **148**, 75–83.

ISO 9308-1 (2014). Water quality – Enumeration of *Escherichia coli* and coliform bacteria – Part 1: Membrane filtration method for waters with low bacterial background flora. https://www.iso.org/obp/ui/#iso:std:iso:9308:-1:ed-3:v1:en (accessed 16 April 2020).

ISO 9308-2 (2012). Water quality – Enumeration of *Escherichia coli* and coliform bacteria – Part 2: Most probable number method. https://www.iso.org/obp/ui/#iso:std: iso:9308:-2:ed-2:v1:en (accessed 16 April 2020).

ISO 9308-3 (1998). Water quality – Detection and enumeration of *Escherichia coli* and coliform bacteria – Part 3: Miniaturized method (Most Probable Number) for the detection and enumeration of *E. coli* in surface and waste water. https://www.iso. org/obp/ui/#iso:std:iso:9308:-3:ed-1:v1:en (accessed 16 April 2020).

Loewenthal M., Newton A. D., Wright S., Campbell C., Crossley A., Hausot A. and Angelescu D. E. (2018). Rapid microbiology field instrumentation: source tracking in sensitive areas. *Institute of Water Magazine*, **2018**(Q3), pp. 86–87.

Lopez-Roldan R., Tusell P., Courtois S. and Cortina J. L. (2013). On-line bacteriological detection in water. *Trends in Analytical Chemistry*, **44**, 46–57.

Mailler R. (2015). Fate of Emerging and Priority Micropollutants in Conventionnal Wastewater Treatement Facilities, Sludge Treatments and Tertiary Treatmeent

Processes Using Activated Carbons. PhD thesis, Université Paris-Est., Paris. https://www.theses.fr/2015PESC1060.

Noble R. T. and Weisberg S. B. (2005) A review of technologies for rapid detection of bacteria in recreational waters. *Journal of Water and Health*, **3**(4), 381–392.

Prüss A. (1998). Review of epidemiological studies on health effects from exposure to recreational water. *International Epidemiological Association*, **27**, 1–9.

Rocher V. and Azimi S. (2016). Microbial Quality of Waters in the Paris Area: From Wastewaters to Surface Waters. Johanet Publisher, Paris. 94 p. ISBN: 979-10-91089-29-6.

US EPA (2014). Site-specific alternative recreational criteria technical support materials for alternative indicators and methods. U.S. Environmental Protection Agency, Office of Water, Document N°. EPA-820-R-14-011.

Wildeboer D., Amirat L., Price R. G. and Abuknesha R. A. (2010). Rapid detection of *Escherichia coli* in water using a hand-held fluorescence detector. *Water Research*, **44**, 2621–2628.

World Health Organization (WHO) (2003). Guidelines for Safe Recreational Water Environments, Chapter 4: Fecal Pollution and Water Quality. World Health Organization, Geneva, Switzerland, pp. 51–101.

Ziegler R. (2019). Viewpoint: water innovation for a circular economy – the contribution of grassroots actors. *Water Alternatives*, **12**(2), 725–738.

Section 2 – Chapter 3

Azzellino A., Antonelli M., Canziani R., Malpei F., Marinetti M. and Nurizzo C. (2011). Multivariate modelling of disinfection kinetics: a comparison among three different disinfectants. *Desalination and Water Treatment*, **29**(1–3), 128–139.

Directive 2006/7/EC of 15 February 2006 concerning the management of bathing water quality and repealing Directive 76/160/EEC, 064.

Domínguez Henao L., Cascio M., Turolla A. and Antonelli M. (2018). Effect of suspended solids on peracetic acid decay and bacterial inactivation kinetics: experimental assessment and definition of predictive models. *Science of the Total Environment*, **643**, 936–945.

Eleria A. and Vogel R. M. (2005). Predicting fecal coliforl bacteria levels in the Charles River, Massachussets, *USA. Journal of the American Water Resources Association*, **41**(5), 1195–1209.

Fernando W. J. N. (2009). Theoretical considerations and modeling of chemical inactivation of microorganisms: inactivation of Giardia Cysts by free chlorine. *Journal of Theoretical Biology*, **259**(2), 297–303.

Flores M. J., Brandi R. J., Cassano A. E. and Labas M. D. (2016). Kinetic model of water disinfection using peracetic acid including synergistic effects. *Water Science and Technology*, **73**(2), 275–282.

Hassen A., Mahrouk M., Ouzari H., Cherif M., Boudabous A. and Damelincourt J. J. (2000) UV disinfection of treated wastewater in a large-scale pilot plant and inactivation of selected bacteria in a laboratory UV device. *Bioresource Technology*, **74**, 141–150.

Herrig I. M., Böer S. I., Brennholt N. and Manz W. (2015) Development of multiple linear regression models as predictive tools for fecal indicator concentrations in a stretch of the lower Lahn River, Germany. *Water Research*, **85**, 148–157.

Karpova T., Pekonen P., Gramstad R., Öjstedt U., Laborda S., Heinonen-Tanski H., Chávez A. and Jiménez B. (2013). Performic acid for advanced wastewater disinfection. *Water Science and Technology*, **68**, 2090–2096. https://doi.org/10.2166/wst.2013.468.

Luukkonen T. and Pehkonen S. O. (2017). Peracids in water treatment: a critical review. *Critical Reviews in Environmental Science and Technology*, **47**, 1–39. https://doi.org/10.1080/10643389.2016.1272343.

Manoli K., Sarathy S., Maffettone R. and Santoro D. (2019). Detailed modeling and advanced control for chemical disinfection of secondary effluent wastewater by peracetic acid. *Water Research*, **153**, 251–262.

Montgomery D. C., Peck E. A. and Vining G. G. (2012) Introduction to Linear Regression Analysis, 5th edn. John Wiley & Sons, Hoboken, NJ.

Murray A., Goldman J., Sarathy S., Hilts B., Bell K., Santoro D., Sun W., Morgan S. and Brower M. (2016). Disinfection of a municipal wastewater secondary effluent with a combination of ultraviolet irradiation and peracetic acid. *Proceedings of the Water Environment Federation*, **2016**(10), 2053–2064.

Pigot T., de Casamajor M. N., Sanchez F., Gonzalez P. and Paulin T. (2019). Disinfection of the treated effluent of the Biarritz wastewater treatment plant using DesinFix: assessment after 38 months. Internal Report.

Ragazzo P., Chiucchini N., Piccolo V. and Ostoich M. (2013). A new disinfection system for wastewater treatment: performic acid full-scale trial evaluations. *Water Science and Technology*, **67**, 2476–2487. https://doi.org/10.2166/wst.2013.137.

Rocher V. and Azimi S. (2016). Microbial Quality of Waters in the Paris Area: From Wastewaters to Surface Waters. Johanet Publisher, Paris. 94 p. ISBN: 979-10-91089-29-6.

Santoro D., Gehr R., Bartrand T. A., Liberti L., Notarnicola M., Dell'Erba A., Falsanisi D. and Haas C. N. (2007). Wastewater disinfection by peracetic acid: assessment of models for tracking residual measurements and inactivation. *Water Environment Research*, **79**(7), 775–787.

Santoro D., Crapulli F., Raisee M., Raspa G. and Haas C. N. (2015). Nondeterministic computational fluid dynamics modeling of *Escherichia coli* inactivation by peracetic acid in municipal wastewater contact tanks. *Environmental Science and Technology*, **49**(12), 7265–7275.

Von Gunten U. (2003). Ozonation of drinking water: part II. Disinfection and by-product formation in presence of bromide, iodide or chlorine. *Water Research*, **37**, 1469–1487. https://doi.org/10.1016/S0043-1354(02)00458-X.

Section 3 – Chapter 1

Alexandrou L., Meehan B. J. and Jones O. A. H. (2018). Regulated and emerging disinfection by-products in recycled waters. *Science of the Total Environment*, **637–638**, 1607–1616.

Ankley G. T., Kahl M. D., Jensen K. M., Hornung M. W., Korte J. J., Makynen E. A. and Leino R. N. (2002). Evaluation of the aromatase inhibitor fadrozole in a short-term reproduction assay with the fathead minnow (Pimephales promelas). *Toxicological Sciences*, **67**(1), 121–130. https://doi.org/10.1093/toxsci/67.1.121.

Becerra-Castro C., Macedo G., Silva A. M. T., Manaia C. M. and Nunes O. C. (2016). Proteobacteria become predominant during regrowth after water disinfection. *Science of the Total Environment*, **573**, 313–323.

Bond T., Huang J., Templeton M. R. and Graham N. (2011). Occurrence and control of nitrogenous disinfection by-products in drinking water – A review. *Water Research*, **45**(15), 4341–4354.

Bouteleux C., Saby S., Tozza D., Cavard J., Lahoussine V., Hartemann P. and Mathieu L. (2005). *Escherichia coli* behavior in the presence of organic matter released by algae exposed to water treatment chemicals. *Applied and Environmental Microbiology*, **71**(2), 734–740.

Chhetri R. K., Thornberg D., Berner J., Gramstad R., Öjstedt U., Sharma A. K. and Andersen H. R. (2014). Chemical disinfection of combined sewer overflow waters using performic acid or peracetic acids. *Science of the Total Environment*, **490**, 1065–1072.

Dell'Erba A., Falsanisi D., Liberti L., Notarnicola M. and Santoro D. (2007). Disinfection by-products formation during wastewater disinfection with peracetic acid. *Desalination*, **215**(1), 177–186.

Domínguez Henao L., Cascio M., Turolla A. and Antonelli M. (2018a). Effect of suspended solids on peracetic acid decay and bacterial inactivation kinetics: experimental assessment and definition of predictive models. *Science of the Total Environment*, **643**, 936–945.

Domínguez Henao L., Delli Compagni R., Turolla A. and Antonelli M. (2018b). Influence of inorganic and organic compounds on the decay of peracetic acid in wastewater disinfection. *Chemical Engineering Journal*, **337**, 133–142.

Du Pasquier D., Lemkine G., Meynerol K., Sauvignet P., Borsato J., Goncalves A. and Rocher V. (2015). Interest of bio-indicator for the performance monitoring of organic micro-pollutants removal from municipal wastewater with dedicated tertiary treatment processes (In French). *Techniques Sciences Méthodes*, **10**, 33–42.

Du Pasquier D., Guérin-Rechdaoui S., Azimi S., Féraudet A., Lemkine G. and Rocher V. (2018). Evolution of endocrine disruption of wastewater during treatment in a WWTP – Use of Watchfrog models. In: To Innovate in Monitoring and Operating Practices for Wastewater Treatment Plants – Scientific and Technical Lessons Learned from Phase I of the Mocopée Program (2014–2017), ASTEE Publisher, Nanterre, pp. 117–127.

Filippis P. D., Scarsella M. and Verdone N. (2009). Peroxyformic acid formation: a kinetic study. *Industrial and Engineering Chemistry Research*, **48**(3), 1372–1375.

Gerrity D., Pisarenko A. N., Marti E., Trenholm R. A., Gerringer F., Reungoat J. and Dickenson E. (2015). Nitrosamines in pilot-scale and full-scale wastewater treatment plants with ozonation. *Water Research*, **72**, 251–261.

Giannakis S., Darakas E., Escalas-Cañellas A. and Pulgarin C. (2015). Environmental considerations on solar disinfection of wastewater and the subsequent bacterial (re) growth. *Photochemical and Photobiological Sciences*, **14**(3), 618–625.

Glover C. M., Verdugo E. M., Trenholm R. A. and Dickenson E. R. V. (2019). N-nitrosomorpholine in potable reuse. *Water Research*, **148**, 306–313.

Heeb M. B., Criquet J., Zimmermann-Steffens S. G. and von Gunten U. (2014). Oxidative treatment of bromide-containing waters: formation of bromine and its reactions with inorganic and organic compounds – A critical review. *Water Research*, **48**, 15–42.

Heinonen-Tanski H. and Miettinen H. (2010). Performic acid as a potential disinfectant at low temperature. *Journal of Food Process Engineering*, **33**(6), 1159–1172.

Held A. M., Halko D. J. and Hurst J. K. (1978). Mechanisms of chlorine oxidation of hydrogen peroxide. *Journal of the American Chemical Society*, **100**(18), 5732–5740.

Jensen K. M., Kahl M. D., Makynen E. A., Korte J. J., Leino R. L., Butterworth B. C. and Ankley G. T. (2004). Characterization of responses to the antiandrogen flutamide in a short-term reproduction assay with the fathead minnow. *Aquatic Toxicology*, **70**(2), 99–110.

Karpova T., Pekonen P., Gramstad R., Öjstedt U., Laborda S., Heinonen-Tanski H., Chávez A. and Jiménez B. (2013). Performic acid for advanced wastewater disinfection. *Water Science and Technology*, **68**(9), 2090–2096.

Katsiadaki I., Morris S., Squires C., Hurst M. R., James J. D. and Scott A. P. (2006). Use of the three-spined stickleback (Gasterosteus aculeatus) as a sensitive in vivo test for the detection of environmental antiandrogens. *Environmental Health Perspectives*, **114** (Suppl. 1), 115–121.

Keefer L. K. and Roller P. P. (1973). N-nitrosation by nitrite ion in neutral and basic medium. *Science*, **181**(4106), 1245–1247.

Kitis M. (2004). Disinfection of wastewater with peracetic acid: a review. *Environment International*, **30**(1), 47–55.

Krauss M., Longrée P., Dorusch F., Ort C. and Hollender J. (2009). Occurrence and removal of N-nitrosamines in wastewater treatment plants. *Water Research*, **43**(17), 4381–4391.

Lee J.-H. and Oh J.-E. (2016). A comprehensive survey on the occurrence and fate of nitrosamines in sewage treatment plants and water environment. *Science of the Total Environment*, **556**, 330–337.

Leloup J. and Buscagalia M. (1977). Triiodothyronine: amphibian metamorphosis hormone. *C. R. Academy of Science*, **284**, 2261–2263.

Le Roux J., Gallard H. and Croué J.-P. (2011). Chloramination of nitrogenous contaminants (pharmaceuticals and pesticides): NDMA and halogenated DBPs formation. *Water Research*, **45**(10), 3164–3174.

Leveneur S., Ledoux A., Estel L., Taouk B. and Salmi T. (2014). Epoxidation of vegetable oils under microwave irradiation. *Chemical Engineering Research and Design*, **92**(8), 1495–1502.

Luukkonen T. and Pehkonen S. O. (2017). Peracids in water treatment: a critical review. *Critical Reviews in Environmental Science and Technology*, **47**(1), 1–39.

Luukkonen T., Heyninck T., Rämö J. and Lassi U. (2015). Comparison of organic peracids in wastewater treatment: disinfection, oxidation and corrosion. *Water Research*, **85**, 275–285.

McFadden M., Loconsole J., Schockling A. J., Nerenberg R. and Pavissich J. P. (2017). Comparing peracetic acid and hypochlorite for disinfection of combined sewer overflows: effects of suspended-solids and pH. *Science of the Total Environment*, **599–600**, 533–539.

Mèche P. (2016). Quality of the Discharged Water from the SIAAP Facilities in 2014 and 2015 with Regard to the Re-Use Regulations. Internal report.

Mengeot M. A., Musu T. and Vogel L. (2016). Endocrine Disruptors: An Occupational Risk in Need of Recognition. Report of the European Trade Union Institute: ETUI (European Trade Union Institute), Brussels.

Mirvish S. S. (1975). Formation of N-nitroso compounds: chemistry, kinetics, and in vivo occurrence. *Toxicology and Applied Pharmacology*, **31**(3), 325–351.

Mitch W. A., Sharp J. O., Trussell R. R., Valentine R. L., Alvarez-Cohen L. and Sedlak D. L. (2003). N-nitrosodimethylamine (NDMA) as a drinking water contaminant: a review. *Environmental Engineering Science*, **20**(5), 389–404.

Mora M., Veijalainen A.-M. and Heinonen-Tanski H. (2018). Performic acid controls better *Clostridium tyrobutyricum* related bacteria than peracetic acid. *Sustainability*, **10**(11), 1–8.

Park S.-H., Wei S., Mizaikoff B., Taylor A. E., Favero C. and Huang C.-H. (2009). Degradation of amine-based water treatment polymers during chloramination as N-nitrosodimethylamine (NDMA) precursors. *Environmental Science and Technology*, **43**(5), 1360–1366.

Pawlowski S., Van Aerle R., Tyler C. R. and Braunbeck T. (2004). Effects of 17a-ethinylestradiol in a fathead minnow (Pimephales promelas) gonadal recrudescence assay. *Ecotoxicology and Environmental Safety*, **57**, 330–345.

Pigot T., de Casamajor M. N., Sanchez F., Gonzalez P. and Paulin T. (2019). Disinfection of the Treated Effluent of the Biarritz Wastewater Treatement Plant Using DesinFix: Assessment After 38 Months. Internal Report.

Ragazzo P., Chiucchini N., Piccolo V. and Ostoich M. (2013). A new disinfection system for wastewater treatment: performic acid full-scale trial evaluations. *Water Science and Technology*, **67**(11), 2476–2487.

Ragazzo P., Feretti D., Monarca S., Dominici L., Ceretti E., Viola G., Piccolo V., Chiucchini N. and Villarini M. (2017). Evaluation of cytotoxicity, genotoxicity, and apoptosis of wastewater before and after disinfection with performic acid. *Water Research*, **116**, 44–52. https://doi.org/10.1016/j.watres.2017.03.016.

Safford H. R. and Bischel H. N. (2019). Flow cytometry applications in water treatment, distribution, and reuse: a review. *Water Research*, **151**, 110–133.

Santacesaria E., Russo V., Tesser R., Turco R. and Di Serio M. (2017). Kinetics of performic acid synthesis and decomposition. *Industrial and Engineering Chemistry Research*, **56**(45), 12940–12952.

Sardana A., Cottrell B., Soulsby D. and Aziz T. N. (2019). Dissolved organic matter processing and photoreactivity in a wastewater treatment constructed wetland. *Science of the Total Environment*, **648**, 923–934.

Schreiber I. M. and Mitch W. A. (2007). Enhanced nitrogenous disinfection by product formation near the breakpoint: implications for nitrification control. *Environmental Science and Technology*, **41**(20), 7039–7046.

Sebire M., Allen Y., Bersuder P. and Katsiadaki I. (2008). The model anti-androgen flutamide suppresses the expression of typical male stickleback reproductive behavior. *Aquatic Toxicology*, **90**, 37–47.

Seki M., Yokota H., Matsubara H., Tsuruda Y., Maeda M., Tadokoro H. and Kobayashi K. M. (2002). Effect of ethinylestradiol on the reproduction and induction of vitellogenin and testis-ova in medaka (Oryzias latipes). *Environmental Toxicology and Chemistry*, **21**(8), 1692–1698.

Shah A. D. and Mitch W. A. (2012). Halonitroalkanes, halonitriles, haloamides, and N-nitrosamines: a critical review of nitrogenous disinfection byproduct formation pathways. *Environmental Science and Technology*, **46**(1), 119–131.

Shah A. D., Liu Z.-Q., Salhi E., Höfer T. and von Gunten U. (2015). Peracetic acid oxidation of saline waters in the absence and presence of H_2O_2: secondary oxidant and disinfection byproduct formation. *Environmental Science and Technology*, **49**(3), 1698–1705.

Shi Y. B., Sachs L. M., Jones P., Li Q. and Ishizuya-Oka A. (1998). Thyroid hormone regulation of *Xenopus laevis* metamorphosis: functions of thyroid hormone receptors and roles of extracellular matrix remodeling. *Wound Repair and Regeneration*, **6**, 314–322.

Sun X., Zhao X., Du W. and Liu D. (2011). Kinetics of formic acid-autocatalyzed preparation of performic acid in aqueous phase. *Chinese Journal of Chemical Engineering*, **19**(6), 964–971.

Tondera K., Klaer K., Koch C., Hamza I. A. and Pinnekamp J. (2016). Reducing pathogens in combined sewer overflows using performic acid. *International Journal of Hygiene and Environmental Health*, **219**(7, Part B), 700–708.

West D. M., Wu Q., Donovan A., Shi H., Ma Y., Jiang H. and Wang J. (2016). N-nitrosamine formation by monochloramine, free chlorine, and peracetic acid disinfection with presence of amine precursors in drinking water system. *Chemosphere*, **153**, 521–527.

Section 3 – Chapter 2

Alexandrou L., Meehan B. J. and Jones O. A. H. (2018). Regulated and emerging disinfection by-products in recycled waters. *Science of the Total Environment*, **637–638**, 1607–1616.

Ankley G. T., Kahl M. D., Jensen K. M., Hornung M. W., Korte J. J., Makynen E. A. and Leino R. N. (2002). Evaluation of the aromatase inhibitor fadrozole in a short-term reproduction assay with the fathead minnow (Pimephales promelas). *Toxicological Sciences*, **67**(1), 121–130. https://doi.org/10.1093/toxsci/67.1.121.

Becerra-Castro C., Macedo G., Silva A. M. T., Manaia C. M. and Nunes O. C. (2016). Proteobacteria become predominant during regrowth after water disinfection. *Science of the Total Environment*, **573**, 313–323.

Bond T., Huang J., Templeton M. R. and Graham N. (2011). Occurrence and control of nitrogenous disinfection by-products in drinking water – a review. *Water Research*, **45**(15), 4341–4354.

Bouteleux C., Saby S., Tozza D., Cavard J., Lahoussine V., Hartemann P. and Mathieu L. (2005). *Escherichia coli* behavior in the presence of organic matter released by algae exposed to water treatment chemicals. *Applied and Environmental Microbiology*, **71**(2), 734–740.

Chhetri R. K., Thornberg D., Berner J., Gramstad R., Öjstedt U., Sharma A. K. and Andersen H. R. (2014). Chemical disinfection of combined sewer overflow waters using performic acid or peracetic acids. *Science of the Total Environment*, **490**, 1065–1072.

Dell'Erba A., Falsanisi D., Liberti L., Notarnicola M. and Santoro D. (2007). Disinfection by-products formation during wastewater disinfection with peracetic acid. *Desalination*, **215**(1), 177–186.

Domínguez Henao L., Cascio M., Turolla A. and Antonelli M. (2018a). Effect of suspended solids on peracetic acid decay and bacterial inactivation kinetics: experimental assessment and definition of predictive models. *Science of the Total Environment*, **643**, 936–945.

Domínguez Henao L., Delli Compagni R., Turolla A. and Antonelli M. (2018b). Influence of inorganic and organic compounds on the decay of peracetic acid in wastewater disinfection. *Chemical Engineering Journal* **337**, 133–142.

Du Pasquier D., Lemkine G., Meynerol K., Sauvignet P., Borsato J., Goncalves A. and Rocher V. (2015). Interest of bio-indicator for the performance monitoring of organic micro-pollutants removal from municipal wastewater with dedicated tertiary treatment processes (In French). *Techniques Sciences Méthodes*, **10**, 33–42.

Du Pasquier D., Guérin-Rechdaoui S., Azimi S., Féraudet A., Lemkine G. and Rocher V. (2018). Evolution of endocrine disruption of wastewater during treatment in a WWTP - Use of Watchfrog models. In: To Innovate in Monitoring and Operating Practices for Wastewater Treatment Plants – Scientific and Technical Lessons Learned from Phase I of the Mocopée Program (2014–2017), ASTEE Publisher, Nanterre, pp. 117–127.

Filippis P. D., Scarsella M. and Verdone N. (2009). Peroxyformic acid formation: a kinetic study. *Industrial and Engineering Chemistry Research*, **48**(3), 1372–1375.

Gerrity D., Pisarenko A. N., Marti E., Trenholm R. A., Gerringer F., Reungoat J. and Dickenson E. (2015). Nitrosamines in pilot-scale and full-scale wastewater treatment plants with ozonation. *Water Research*, **72**, 251–261.

Giannakis S., Darakas E., Escalas-Cañellas A. and Pulgarin C. (2015). Environmental considerations on solar disinfection of wastewater and the subsequent bacterial (re) growth. *Photochemical and Photobiological Sciences*, **14**(3), 618–625.

Glover C. M., Verdugo E. M., Trenholm R. A. and Dickenson E. R. V. (2019). N-nitrosomorpholine in potable reuse. *Water Research*, **148**, 306–313.

Heeb M. B., Criquet J., Zimmermann-Steffens S. G. and von Gunten U. (2014). Oxidative treatment of bromide-containing waters: formation of bromine and its reactions with inorganic and organic compounds – A critical review. *Water Research*, **48**, 15–42.

Heinonen-Tanski H. and Miettinen H. (2010). Performic acid as a potential disinfectant at low temperature. *Journal of Food Process Engineering*, **33**(6), 1159–1172.

Held A. M., Halko D. J. and Hurst J. K. (1978). Mechanisms of chlorine oxidation of hydrogen peroxide. *Journal of the American Chemical Society*, **100**(18), 5732–5740.

Jensen K. M., Kahl M. D., Makynen E. A., Korte J. J., Leino R. L., Butterworth B. C. and Ankley G. T. (2004). Characterization of responses to the antiandrogen flutamide in a short-term reproduction assay with the fathead minnow. *Aquatic Toxicology*, **70**(2), 99–110.

Karpova T., Pekonen P., Gramstad R., Öjstedt U., Laborda S., Heinonen-Tanski H., Chávez A. and Jiménez B. (2013). Performic acid for advanced wastewater disinfection. *Water Science and Technology*, **68**(9), 2090–2096.

Katsiadaki I., Morris S., Squires C., Hurst M. R., James J. D. and Scott A. P. (2006). Use of the three-spined stickleback (Gasterosteus aculeatus) as a sensitive in vivo test for the detection of environmental antiandrogens. *Environmental Health Perspectives*, **114** (Suppl. 1), 115–121.

Keefer L. K. and Roller P. P. (1973). N-nitrosation by nitrite ion in neutral and basic medium. *Science*, **181**(4106), 1245–1247.

Kitis M. (2004). Disinfection of wastewater with peracetic acid: a review. *Environment International*, **30**(1), 47–55.

Krauss M., Longrée P., Dorusch F., Ort C. and Hollender J. (2009). Occurrence and removal of N-nitrosamines in wastewater treatment plants. *Water Research*, **43**(17), 4381–4391.

Lee J.-H. and Oh J.-E. (2016). A comprehensive survey on the occurrence and fate of nitrosamines in sewage treatment plants and water environment. *Science of the Total Environment*, **556**, 330–337.

Leloup J. and Buscagalia M. (1977). Triiodothyronine: Amphibian metamorphosis hormone. *Czech Republic Academy of Science*, **284**, 2261–2263.

Le Roux J., Gallard H. and Croué J.-P. (2011). Chloramination of nitrogenous contaminants (pharmaceuticals and pesticides): NDMA and halogenated DBPs formation. *Water Research*, **45**(10), 3164–3174.

Leveneur S., Ledoux A., Estel L., Taouk B. and Salmi T. (2014). Epoxidation of vegetable oils under microwave irradiation. *Chemical Engineering Research and Design*, **92**(8), 1495–1502.

Luukkonen T. and Pehkonen S. O. (2017). Peracids in water treatment: a critical review. *Critical Reviews in Environmental Science and Technology*, **47**(1), 1–39.

Luukkonen T., Heyninck T., Rämö J. and Lassi U. (2015). Comparison of organic peracids in wastewater treatment: disinfection, oxidation and corrosion. *Water Research*, **85**, 275–285.

McFadden M., Loconsole J., Schockling A. J., Nerenberg R. and Pavissich J. P. (2017). Comparing peracetic acid and hypochlorite for disinfection of combined sewer overflows: effects of suspended-solids and pH. *Science of the Total Environment*, **599–600**, 533–539.

Mèche P. (2016). Quality of the Discharged Water from the SIAAP Facilities in 2014 and 2015 with Regard to the Re-Use Regulations. Internal report.

Mengeot M. A., Musu T. and Vogel L. (2016). Endocrine Disruptors: An Occupational Risk in Need of Recognition. Report of the European Trade Union Institute. ETUI (European Trade Union Institute), Brussels.

Mirvish S. S. (1975). Formation of N-nitroso compounds: chemistry, kinetics, and in vivo occurrence. *Toxicology and Applied Pharmacology*, **31**(3), 325–351.

Mitch W. A., Sharp J. O., Trussell R. R., Valentine R. L., Alvarez-Cohen L. and Sedlak D. L. (2003). N-nitrosodimethylamine (NDMA) as a drinking water contaminant: a review. *Environmental Engineering Science*, **20**(5), 389–404.

Mora M., Veijalainen A.-M. and Heinonen-Tanski H. (2018). Performic acid controls better *Clostridium tyrobutyricum* related bacteria than peracetic acid. *Sustainability*, **10**(11), 1–8.

Park S.-H., Wei S., Mizaikoff B., Taylor A. E., Favero C. and Huang C.-H. (2009). Degradation of amine-based water treatment polymers during chloramination as N-nitrosodimethylamine (NDMA) precursors. *Environmental Science and Technology*, **43**(5), 1360–1366.

Pawlowski S., Van Aerle R., Tyler C. R. and Braunbeck T. (2004). Effects of 17a-ethinylestradiol in a fathead minnow (Pimephales promelas) gonadal recrudescence assay. *Ecotoxicology and Environmental Safety*, **57**, 330–345.

Pigot T., de Casamajor M. N., Sanchez F., Gonzalez P. and Paulin T. (2019). Disinfection of the Treated Effluent of the Biarritz Wastewater Treatement Plant Using DesinFix: Assessment After 38 Months. Internal Report.

Ragazzo P., Chiucchini N., Piccolo V. and Ostoich M. (2013). A new disinfection system for wastewater treatment: performic acid full-scale trial evaluations. *Water Science and Technology*, **67**(11), 2476–2487.

Ragazzo P., Feretti D., Monarca S., Dominici L., Ceretti E., Viola G., Piccolo V., Chiucchini N. and Villarini M. (2017). Evaluation of cytotoxicity, genotoxicity, and apoptosis of wastewater before and after disinfection with performic acid. *Water Research*, **116**, 44–52. https://doi.org/10.1016/j.watres.2017.03.016.

Safford H. R. and Bischel H. N. (2019). Flow cytometry applications in water treatment, distribution, and reuse: a review. *Water Research*, **151**, 110–133.

Santacesaria E., Russo V., Tesser R., Turco R. and Di Serio M. (2017). Kinetics of performic acid synthesis and decomposition. *Industrial and Engineering Chemistry Research*, **56**(45), 12940–12952.

Sardana A., Cottrell B., Soulsby D. and Aziz T. N. (2019). Dissolved organic matter processing and photoreactivity in a wastewater treatment constructed wetland. *Science of the Total Environment*, **648**, 923–934.

Schreiber I. M. and Mitch W. A. (2007). Enhanced nitrogenous disinfection by product formation near the breakpoint: implications for nitrification control. *Environmental Science and Technology*, **41**(20), 7039–7046.

Sebire M., Allen Y., Bersuder P. and Katsiadaki I. (2008). The model anti-androgen flutamide suppresses the expression of typical male stickleback reproductive behavior. *Aquatic Toxicology*, **90**, 37–47.

Seki M., Yokota H., Matsubara H., Tsuruda Y., Maeda M., Tadokoro H. and Kobayashi K. M. (2002). Effect of ethinylestradiol on the reproduction and induction of vitellogenin and testis-ova in medaka (Oryzias latipes). *Environmental Toxicology and Chemistry*, **21**(8), 1692–1698.

Shah A. D. and Mitch W. A. (2012). Halonitroalkanes, halonitriles, haloamides, and N-nitrosamines: a critical review of nitrogenous disinfection byproduct formation pathways. *Environmental Science and Technology*, **46**(1), 119–131.

Shah A. D., Liu Z.-Q., Salhi E., Höfer T. and von Gunten U. (2015). Peracetic acid oxidation of saline waters in the absence and presence of H_2O_2: secondary oxidant and disinfection byproduct formation. *Environmental Science and Technology*, **49**(3), 1698–1705.

Shi Y. B., Sachs L. M., Jones P., Li Q. and Ishizuya-Oka A. (1998). Thyroid hormone regulation of Xenopus laevis metamorphosis: functions of thyroid hormone receptors and roles of extracellular matrix remodeling. *Wound Repair and Regeneration*, **6**, 314–322.

Sun X., Zhao X., Du W. and Liu D. (2011). Kinetics of formic acid-autocatalyzed preparation of performic acid in aqueous phase. *Chinese Journal of Chemical Engineering*, **19**(6), 964–971.

Tondera K., Klaer K., Koch C., Hamza I. A. and Pinnekamp J. (2016). Reducing pathogens in combined sewer overflows using performic acid. *International Journal of Hygiene and Environmental Health*, **219**(7, Part B), 700–708.

West D. M., Wu Q., Donovan A., Shi H., Ma Y., Jiang H. and Wang J. (2016). N-nitrosamine formation by monochloramine, free chlorine, and peracetic acid disinfection with presence of amine precursors in drinking water system. *Chemosphere*, **153**, 521–527.

Section 4 – Chapter 1

Andral B., Boissery P., Descamp P. and Guilbert A. (2011). Monitoring of Urban Discharges and Sanitation Systems in the Mediterranean Sea. Report of the Rhône-Méditeranée Corse River Bassin Organization, 2nd édition, L'OEil d'Andromède edition, Carnon, France.

Bachelot M. (2012). Organic UV filter concentration in marine mussels from french coastal regions. *Science of the Total Environment*, **420**, 273–279. https://doi.org/10.1016/j.scitotenv.2011.12.051.

Cabral-Oliveira J., Dolbeth M. and Pardal A. (2014). Impact of seawage pollution on the structure and functioning of a rocky shore benthic community. *Marine and Freshwater Research*, **65**, 750–758. https://doi.org/10.1071/MF13190.

de Casamajor M.-N. (2004). Bay of Biscay: Lack of Knowledge and Diversity. Alexandre Dewez edn, Ascain, France.

De los Ríos A., Juanes J. A., Ortiz-Zarragoitia M., López de Alda M., Barceló D. and Cajaraville M. P. (2012). Assessment of the effects of a marine urban outfall discharge on caged mussels using chemical and biomarker analysis. *Marine Pollution Bulletin*, **64**, 563–573. https://doi.org/10.1016/j.marpolbul.2011.12.018.

Gosling E. (2003). Bivalve Molluscs – Biology, Ecology and Culture. Fishing News Books. Blackwell Publishing Ltd., Oxford.

Kerambrun E., Henry F., Sanchez W. and Amara R. (2012). Relationships between biochemical and physiological biomarkers responses measured on juvenile marine fish under environmental chemical contamination. *Comparative Biochemistry and Physiology A: Molecular and Integrative Physiology*, **163**, 522–523. https://doi.org/10.1016/j.cbpa.2012.05.071.

Ragazzo P., Chiucchini N., Piccolo V. and Ostoich M. (2013). A new disinfection system for wastewater treatment: performic acid full-scale trial evaluations. *Water Science and Technology*, **67**, 2476–2487. https://doi.org/10.2166/wst.2013.137.

Seed R. and Suchanek T. H. (1992). Population and community ecology of *Mytilus*. In: The mussel Mytilus: Ecology, Physiology, Genetic and Culture, E. M. Gosling (ed), Elsevier Science Publishers, Amsterdam, the Netherlands, pp. 87–169.

Swiacka K., Maculewicz J., Smolarz K., Szaniawaska A. and Caban M. (2019). Mytilidae as model organisms in the marine ecotoxicology of pharmaceuticals – a review. *Internal Pollution*, **254**, 113082. https://doi.org/10.1016/j.envpol.2019.113082.

Terlizzi A., Fraschetti S., Guidetti P. and Boero F. (2002). The effects of sewage discharge on shallow hard substrate sessile assemblages. *Marine Pollution Bulletin*, **44**, 544–550. https://doi.org/10.1016/s0025-326x(01)00282-x.

Turja R., Lehtonen K. K., Meierjohann A., Brozinski J.-M., Vahtera E., Soirinsuo A., Sokolov A., Snoeijs P., Budzinski H., Devier M.-H., Peluhet L., Pääkkönen J.-P., Viitasalo M. and Kronberg L. (2015). The mussel caging approach in assessing biological effects of wastewater treatment plant discharges in the Gulf of Finland (Baltic Sea). *Marine Pollution Bulletin*, **97**, 135–149. https://doi.org/10.1016/j.marpolbul.2015.06.024.

Walne P. R. and Mann R. (1975). Growth and biochemical composition in Ostrea edulis and Crassostrea gigas. In: Ninth European Marine Biology Symposium, Aberdeen University Press, Scotland, UK, pp. 587–607.

Williams G. J. (2011). Data Mining with Rattle and R: The Art of Excavating Data for Knowledge Discovery. Springer Science and Business Media, New York.

Section 4 – Chapter 2

APAT CNR-IRSA Man 29 (2003). Analytical Methods for Water. Report 29/2003. APAT, Roma, Italia.

APHA, AWWA, WEF (2012). Standard Methods for the Examination of Water and Wastewater. 22nd edn. American Public Health Association, Washington DC.

Gagnon C., Lajeunesse A., Cejka P., Gagné F. and Hausler R. (2008). Degradation of selected acidic and neutral pharmaceutical products in a primary-treated wastewater by disinfection processes. *Ozone: Science and Engineering*, **30**(5), 387–397. https://dx.doi.org/10.1080/01919510802336731.

Greenspan F. P. and and Mackellar D. G. (1948). Analysis of aliphatic per acids. *Analytical chemistry*, **20**(11), 1061–1063.

Luukkonen T. and Pehkonen S. O. (2017). Peracids in water treatment: a critical review. *Critical Reviews in Environmental Science and Technology*, **47**(1), 1–39. https://doi.org/10.1080/10643389.2016.1272343.

Luukkonen T., Teeriniemi J., Prokkola H., Ramo J. and Lassi U. (2014). Chemical aspects of peracetic acid based wastewater disinfection. *Water SA*, **40**(1), 73–80. https://dx.doi.org/10.4314/wsa.v40i1.9.

Luukkonen T., Heyninck T., Rämö J. and Lassi U. (2015). Comparison of organic peracids in wastewater treatment: disinfection, oxidation and corrosion. *Water Research*, **85**, 275–285. https://dx.doi.org/10.1016/j.watres.2015.08.037.

Maffettone R., Sarathy S., Wen Y., Passalacqua K., Neofotistos P., Wobus C. and Santoro D. (2018). Techno-economic Evaluation of Alternative Chemical Disinfectants and UV For Wastewater Disinfection: inactivation of bacterial indicator, F-specific and somatic coliphages and MNV. EcoSTP2018 – Ecotechnologies for Wastewater Treatment, London, Canada.

Ragazzo P., Chiucchini N. and Bottin F. (2007). The use of hyproform disinfection system in wastewater treatment: batch and full-scale trials. In: Chemical Water and Wastewater Treatment IX, H. H. Hahn, E. Hoffmann and H. Odegaard (eds), IWA Publishing, London, UK, pp. 267–275.

Ragazzo P., Chiucchini N., Piccolo V. and Ostoich M. (2013). A new disinfection system for wastewater treatment: performic acid full-scale trial evaluations. *Water Science and Technology*, **67**(11), 2476–2487. https://dx.doi.org/10.2166/wst.2013.137.

Ragazzo P., Feretti D., Monarca S., Dominici L., Ceretti E., Viola G., Piccolo V., Chiucchini N. and Villarini M. (2017). Evaluation of cytotoxicity, genotoxicity, and apoptosis of wastewater before and after disinfection with performic acid. *Water Research*, **116**, 44–52. https://doi.org/10.1016/j.watres.2017.03.016.

Ragazzo P., Chiucchini N., Piccolo V., Carrer S., Zanon F. and Gehr R. (2020). Wastewater disinfection: long-term laboratory and full-scale studies on performic acid in comparison with peracetic acid and chlorine. *Water Research*, **184**, 116169. doi: 10.1016/j.watres.2020.116169. Epub 2020 Jul 11. PMID: 32707309.

Cultural aversion to microbes, healthiness or desire for safe bathing, the applications for water disinfection are varied and the technologies used to achieve this goal are numerous. The authors looked at a simple solution to implement: the use of a reagent called performic acid. Consequently, more than two years of applied research, observations and analyzes were necessary to demonstrate its harmlessness towards the natural environment. The strength of the demonstration lies in the cross-vision of many researchers and scientists from different backgrounds who shared their studies and observations. The strength of this testimony also lies in the diversity of the application cases, including notable and sensitive receiving environments as different as the Seine, the Atlantic Ocean or the Venice lagoon. Through its intentions and results, this work is a step moving forward the 2030 Agenda for Sustainable Development, particularly SDG 6 'Clean water and sanitation' relying on the lever of SDG 17 'Partnerships for the goals'.

Denis Penouel
Deputy Chief Executive Officer in charge of prospective

List of Figures

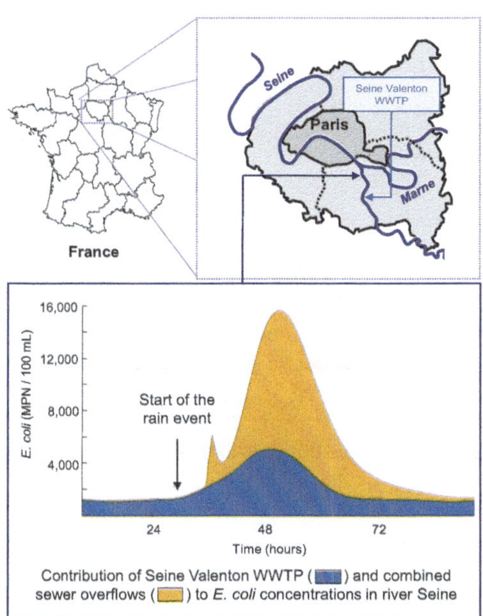

Figure 1 Contribution of both the Seine Valenton WWTP discharge (blue curve) and the combined sewer overflows (yellow curve) to bacterial contamination (*E. coli*) of the Seine River, as simulated using the PROSE mathematical model at Port à l'anglais site, located downstream of the plant.

SECTION 1 – CHAPTER 1

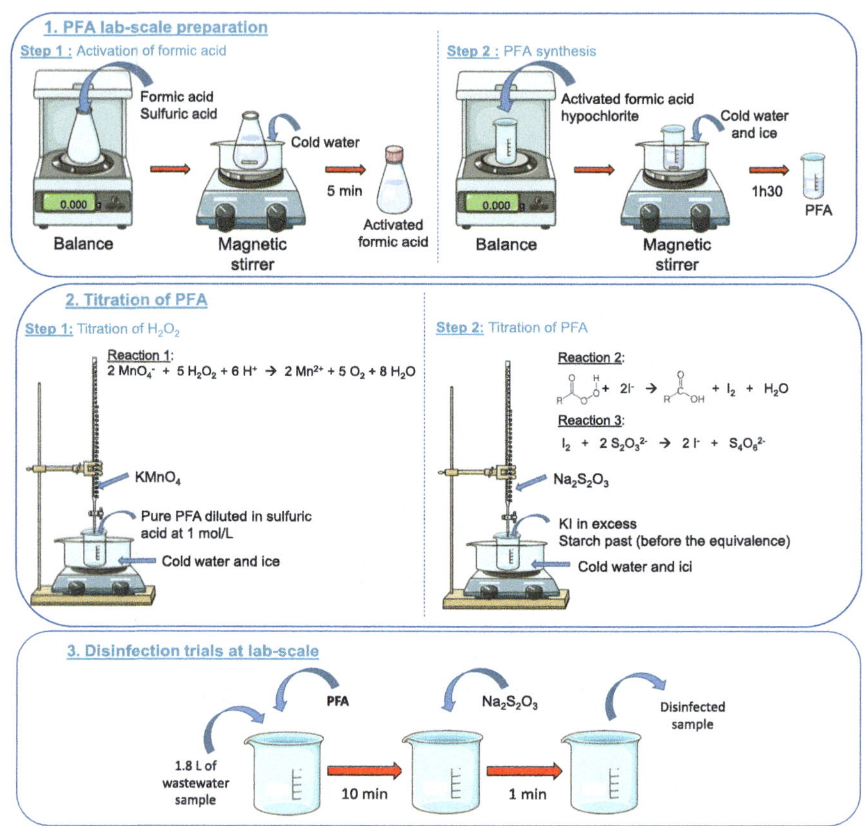

Figure 2 Summary of PFA preparation, titration and laboratory-scale disinfection trial methods.

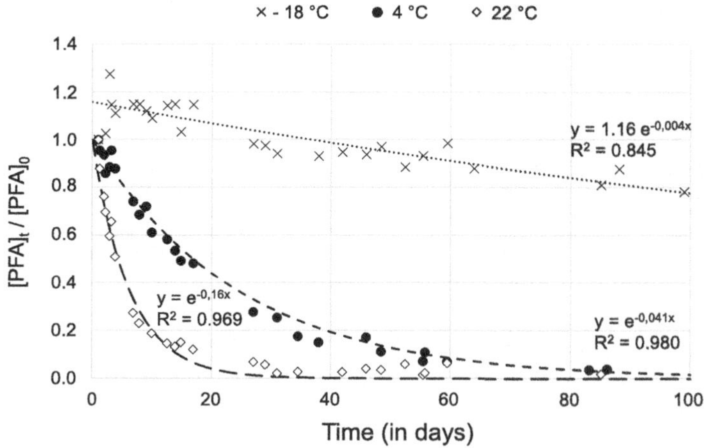

Figure 3 Decay kinetics of PFA produced at the laboratory scale and stored under various conditions.

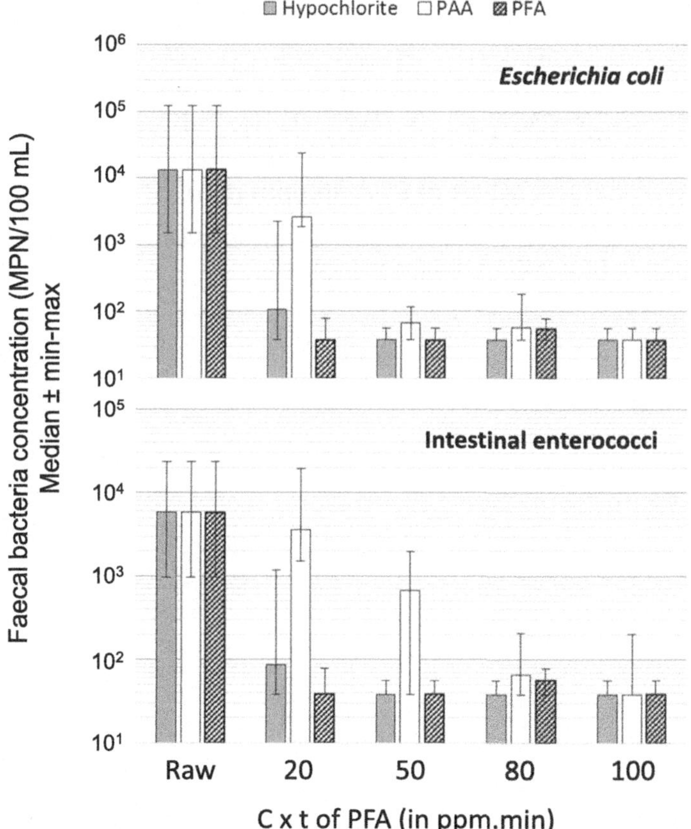

Figure 4 Batch scale comparison of fecal bacteria disinfection effectiveness of PFA, PAA and hypochlorite applied to SEV WWTP discharge (contact time: 10 min).

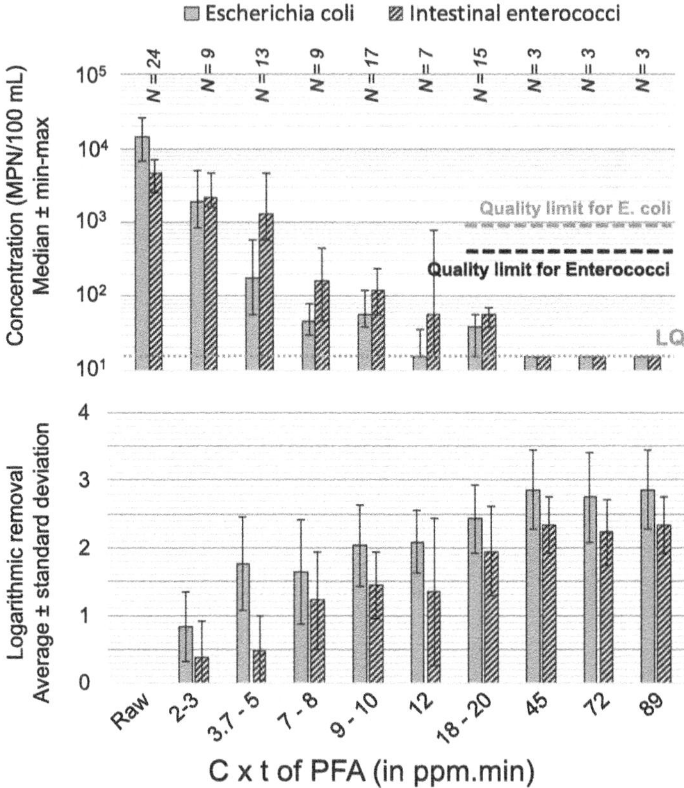

Figure 5 Batch-scale assessment of fecal bacteria disinfection effectiveness by PFA applied to multiple SEV WWTP discharge samples.

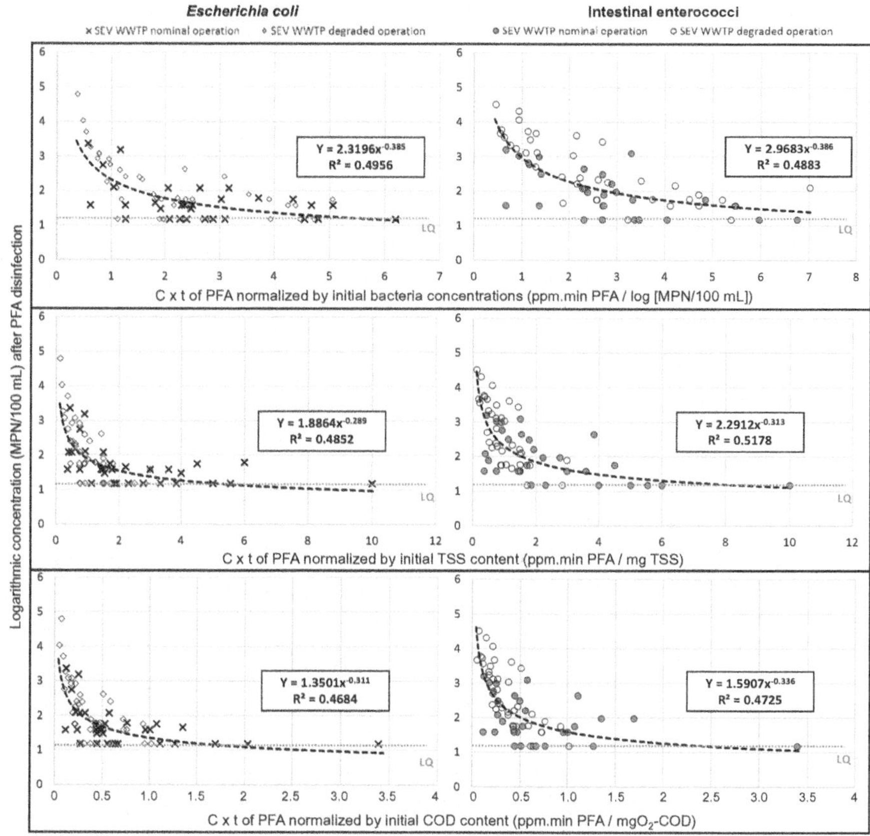

Figure 6 Evaluation of PFA disinfection effectiveness by normalizing the applied PFA dose to the initial bacterial concentrations, TSS and COD for a contact time of 10 min.

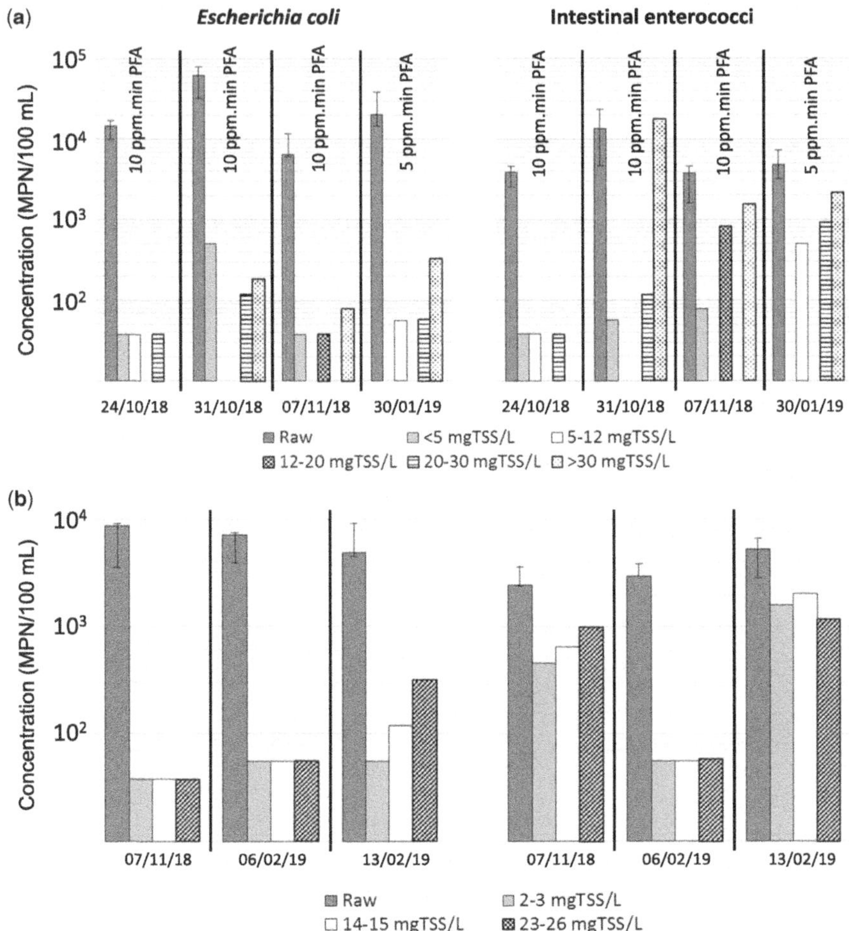

Figure 7 Batch-scale assessment of TSS concentration impact on PFA disinfection effectiveness in SEV WWTP discharge – (a) spiked with raw TSS and (b) spiked with autoclaved TSS.

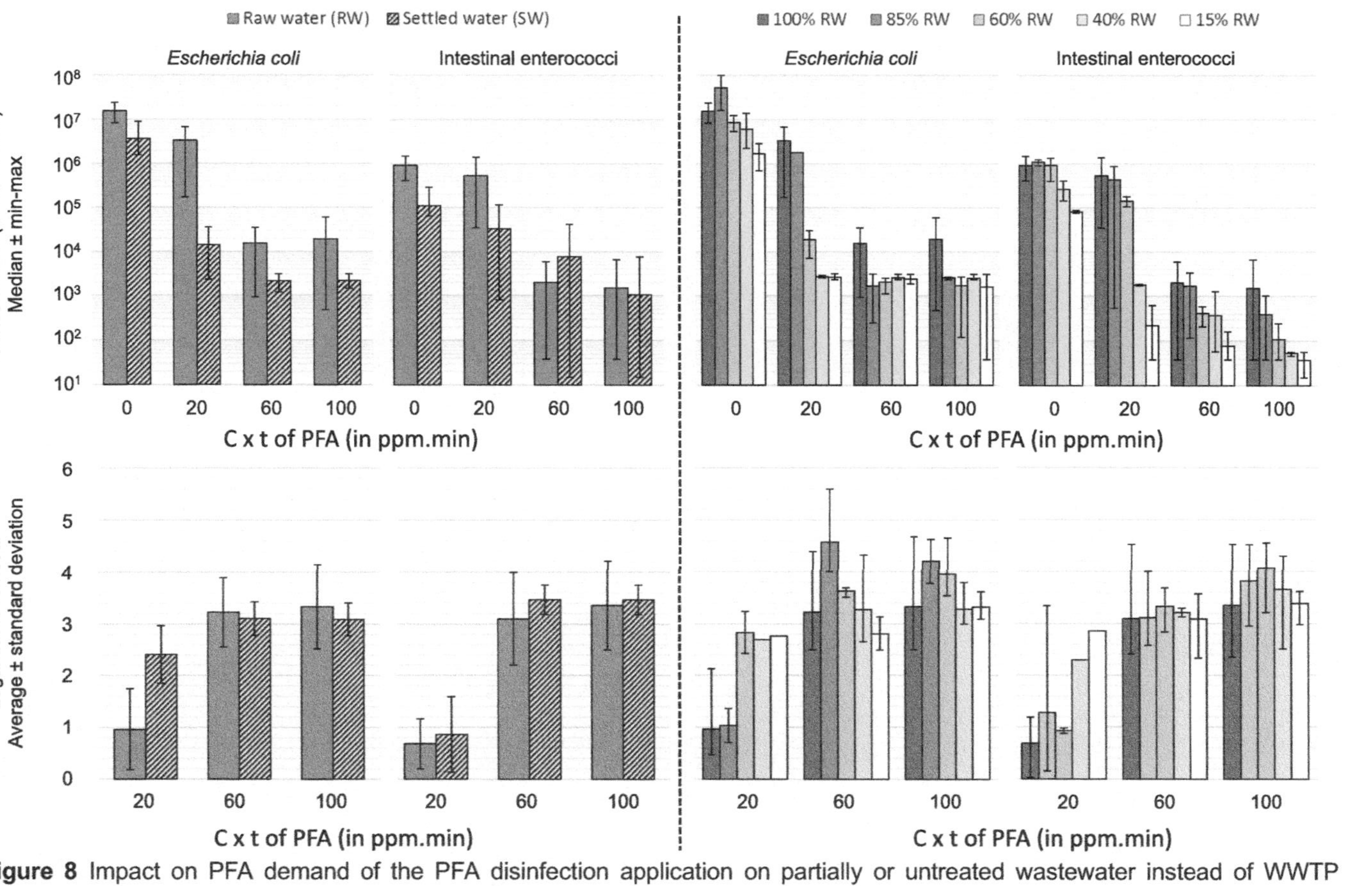

Figure 8 Impact on PFA demand of the PFA disinfection application on partially or untreated wastewater instead of WWTP discharge.

SECTION 1 – CHAPTER 2

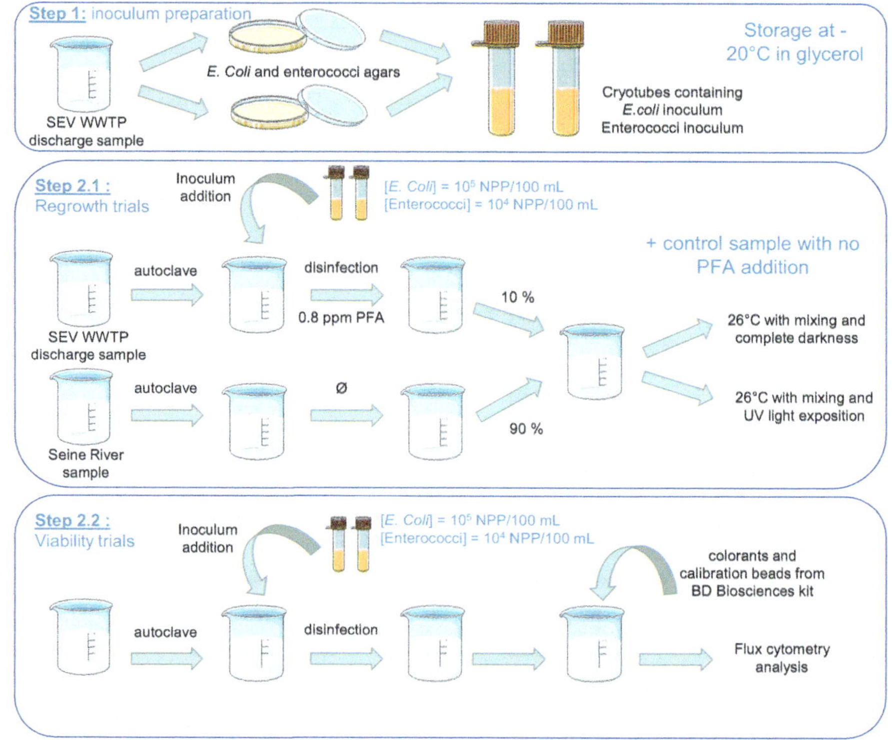

Figure 9 Experimental procedure applied during regrowth and viability trials.

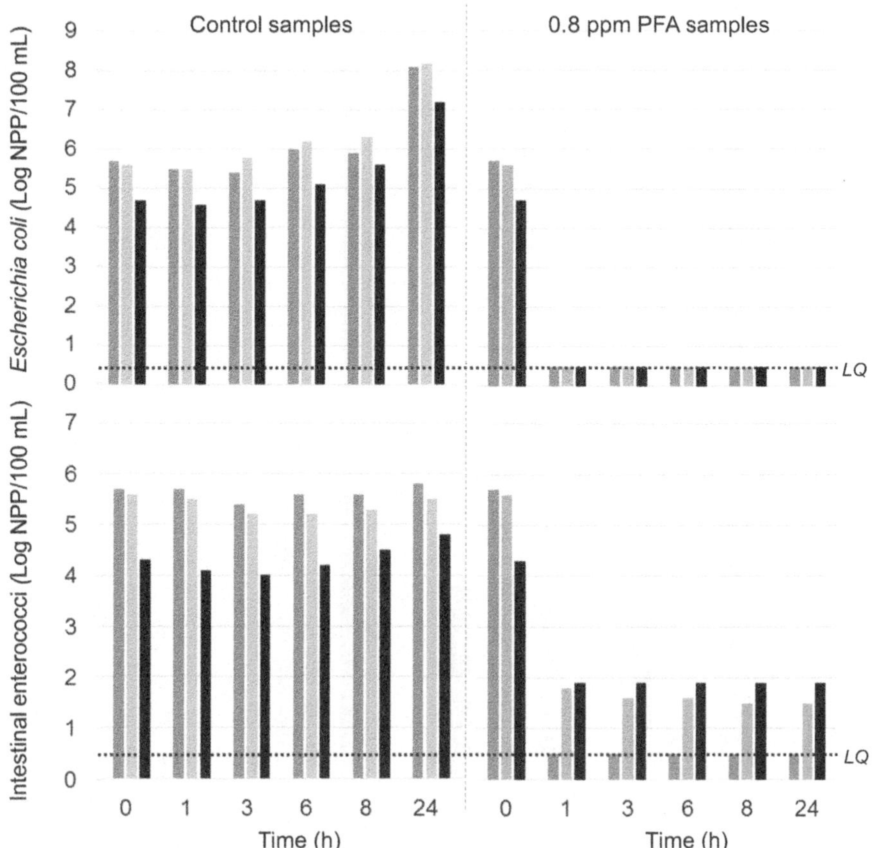

Figure 10 Batch evaluation of fecal bacteria regrowth in a wastewater and seine river composite sample after disinfection with 0.8 ppm of PFA.

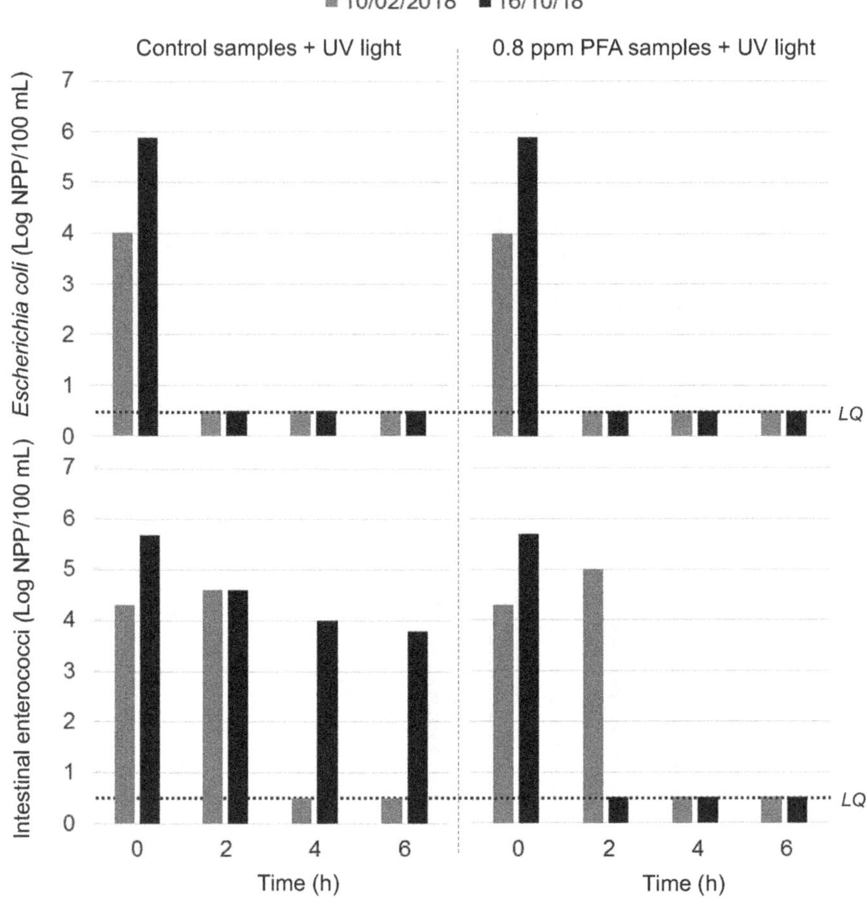

Figure 11 Impact of UV light exposure on fecal bacteria regrowth in a wastewater and seine river composite sample after disinfection with 0.8 ppm of PFA.

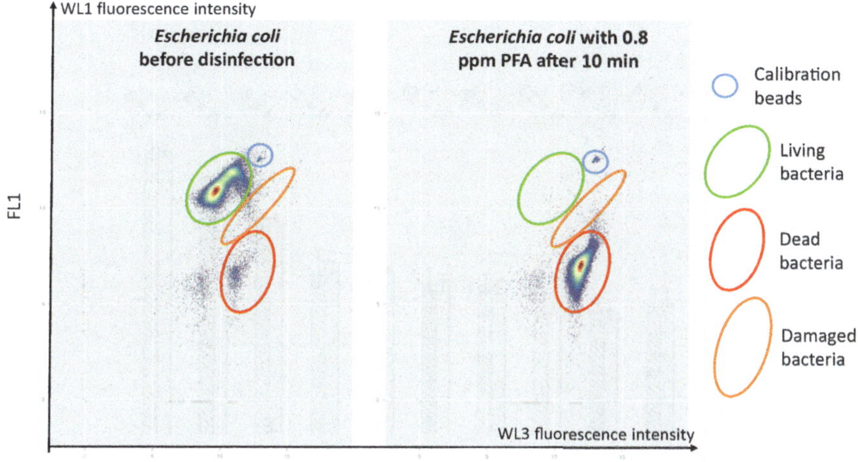

Figure 12 Results of the flux cytometry analysis both before and after disinfection of the spiked *E. coli* solution with 0.8 ppm of PFA after 10 min.

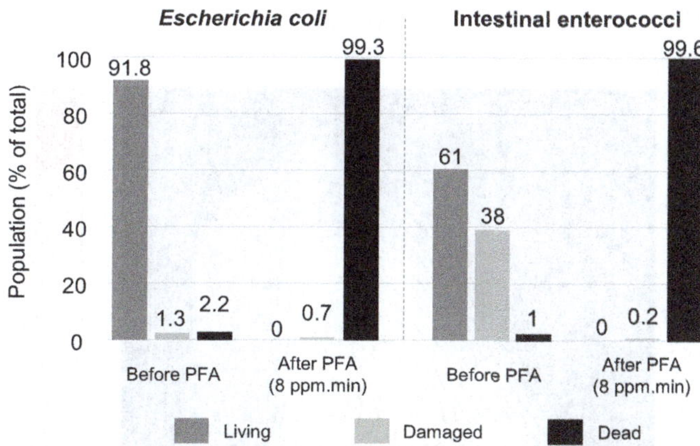

Figure 13 Quantification of the fecal bacteria distribution before and after PFA disinfection (0.8 ppm for 10 min) based on cytometry analysis.

SECTION 1 – CHAPTER 3

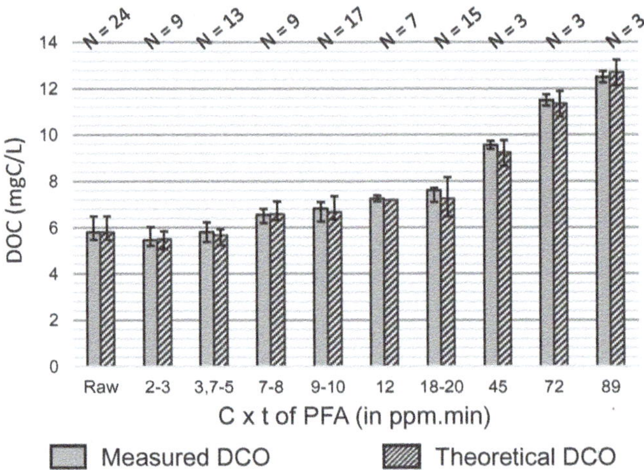

Figure 14 Impact of PFA injection at various C × t on the dissolved organic carbon content of SEV WWTP discharge samples.

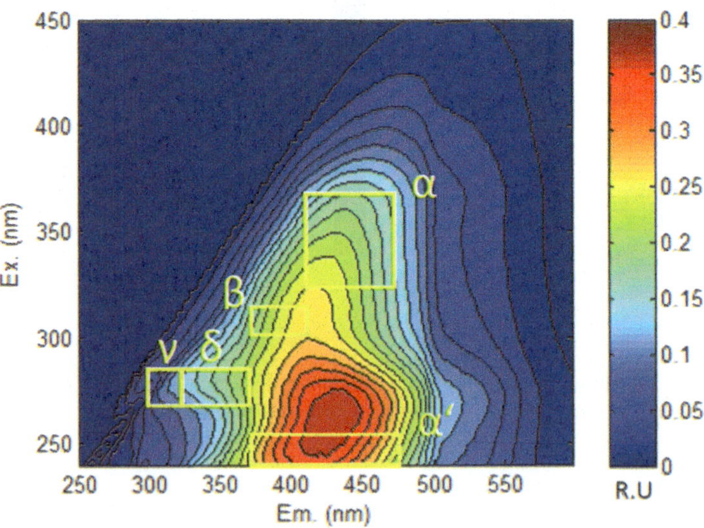

Figure 15 3D fluorescence spectrum example with localization of DOM fluorophore areas of interest.

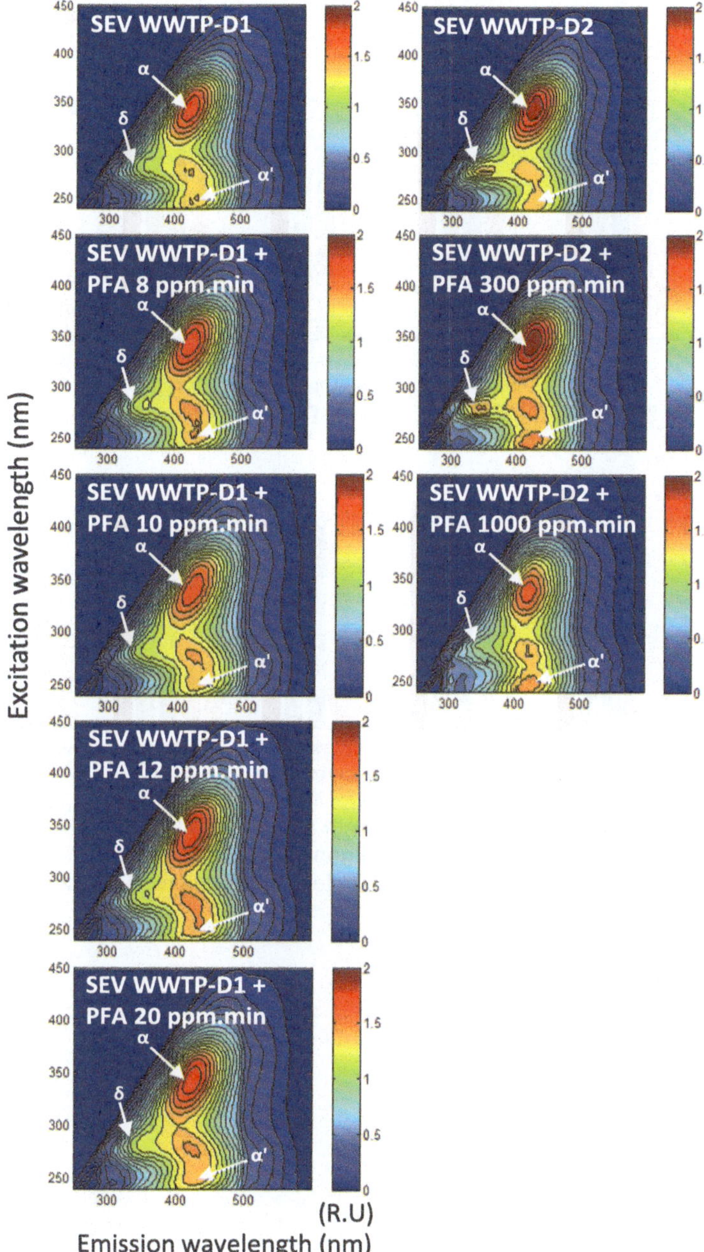

Figure 16 3D EEM spectra obtained for two SEV WWTP discharge samples (D1: sept 4, 2018; D2: Nov 6, 2018) before and after disinfection at various PFA concentrations.

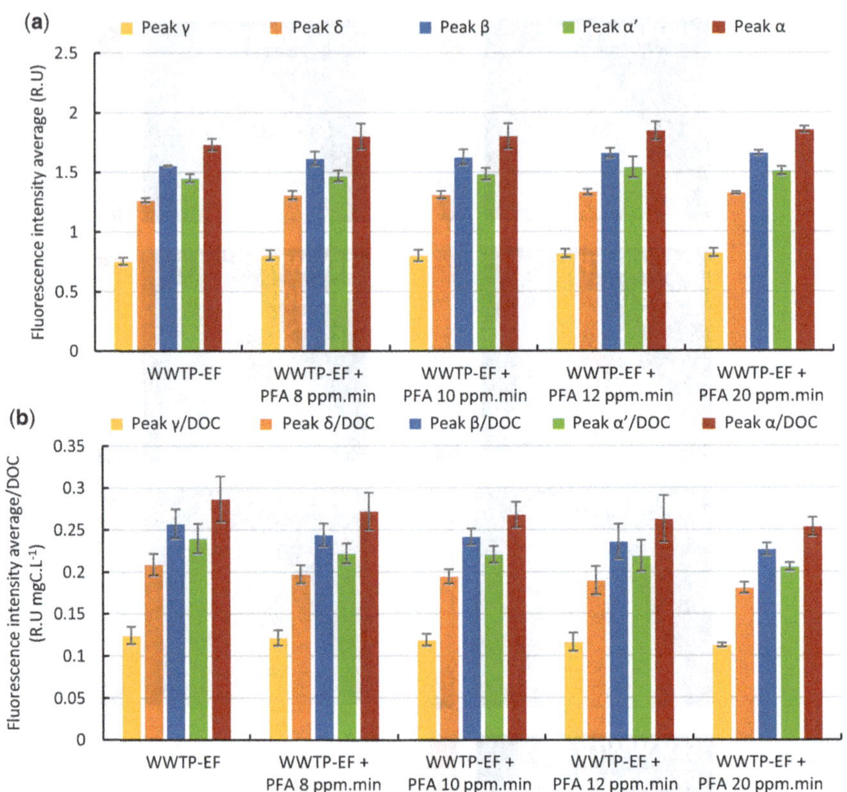

Figure 17 Average fluorescence intensities ($n = 3$) (a) not standardized to DOC, and (b) standardized to the DOC of dissolved organic matter fluorophores in SEV WWTP discharges; (D) non-disinfected and disinfected at operational PFA concentrations (8, 10, 12 and 20 ppm.min).

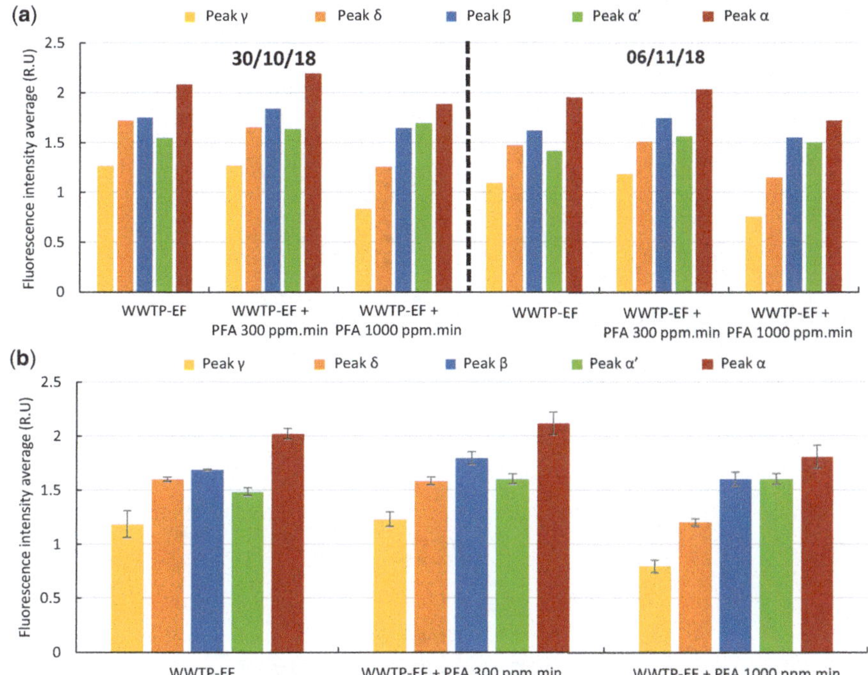

Figure 18 (a) Fluorescence intensities (Oct 30 and Nov 6, 2018) and (b) average ($n = 2$) fluorescence intensities of dissolved organic matter fluorophores in SEV WWTP discharge; (D) before and after disinfection tests at 300 and 1000 ppm.min of PFA.

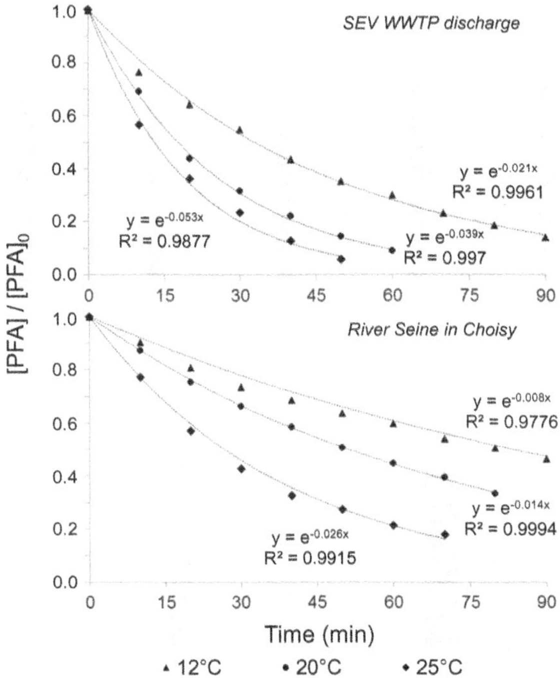

Figure 19 PFA degradation kinetics observed in the autoclaved samples (sept 11, 2018) of SEV WWTP discharge and seine river water in choisy spiked with 2.5 ppm of PFA.

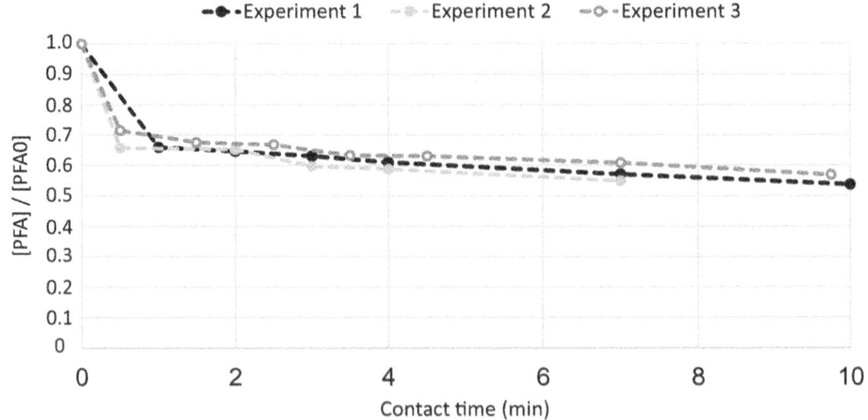

Figure 20 PFA degradation kinetics observed in raw samples (Oct 1, 2019, Oct 29, 2019 and Nov 5, 2019) of SEV WWTP discharge spiked with 1–1.2 ppm of PFA.

SECTION 1 – CHAPTER 4

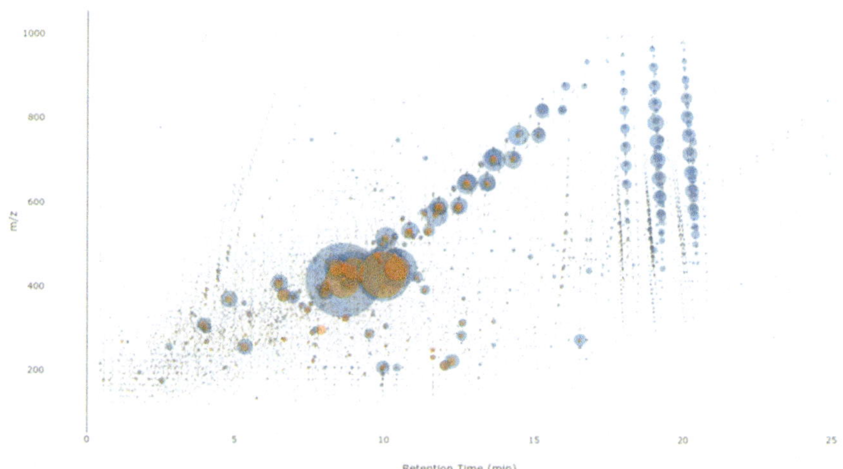

Figure 21 Organic compound fingerprints obtained by high-resolution mass spectrometry (UPLC-IMS-QTOF) for treated water from the seine amont valenton WWTP (SEV-TW, Nov 6, 2018) before (blue) and after (orange) disinfection with 2 ppm of PFA for 10 min. The width of circles is proportional to the marker intensity.

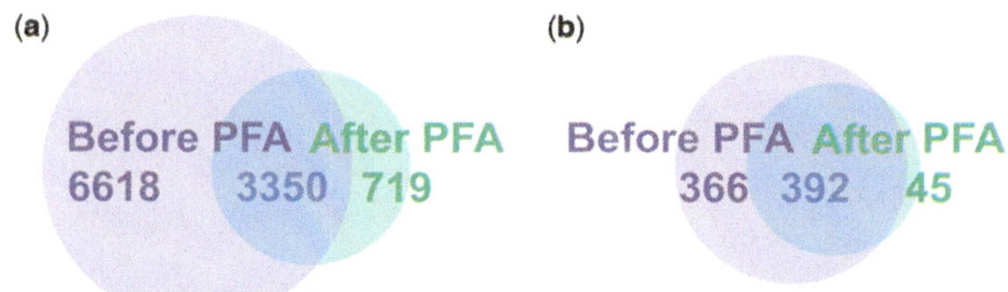

Figure 22 Euler diagrams describing the number of markers with intensities above 10,000 detected in: (a) positive ionization mode and (b) negative ionization mode by UPLC-IMS-QTOF analysis from WWTP discharge (Nov 6, 2018) before (purple) and after (green) disinfection with 2 ppm of PFA for 10 min, along with the number of common markers present in both samples before and after PFA addition. The areas are proportional to the number of markers in each zone.

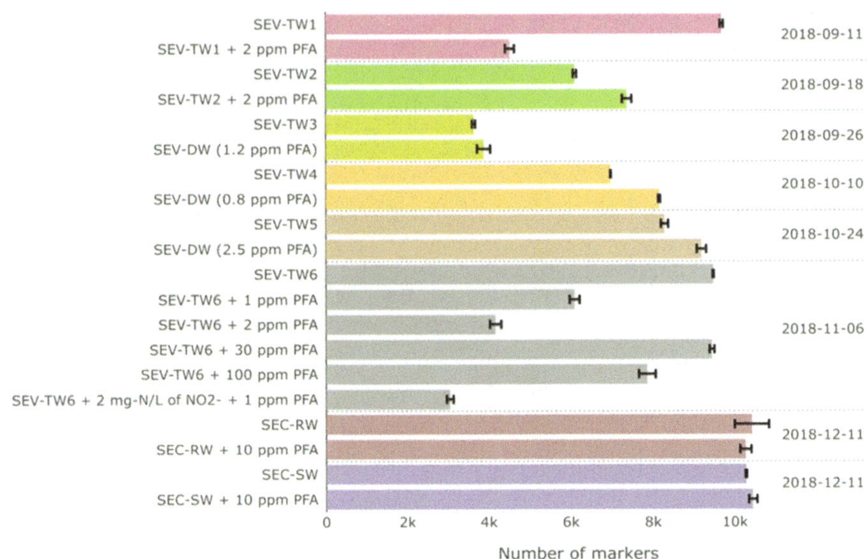

Figure 23 Total number of markers with intensities above 10,000 detected in positive ionization mode by UPLC-IMS-QTOF analysis conducted on all samples before and after PFA disinfection. SEV: seine amont valenton WWTP, SEC: seine centre WWTP, TW: treated water (discharge), DW: disinfected water at the full scale, RW: raw wastewater, SW: settled water. error bars represent the standard deviation from triplicate injections.

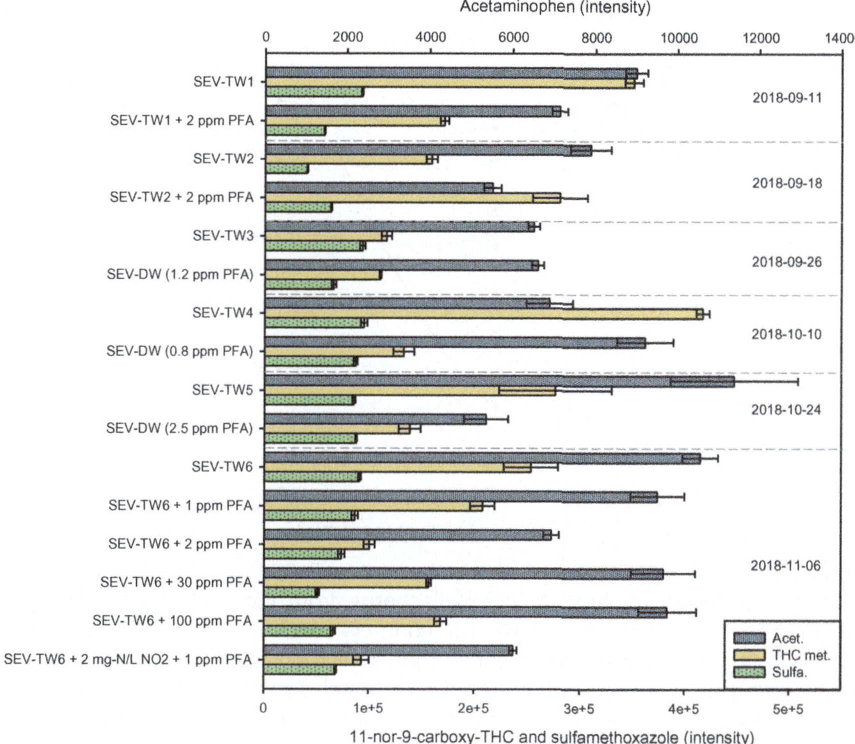

Figure 24 Occurrence of Acetaminophen, 11-nor-9-carboxy-tetrahydrocannabinol (THC metabolite) and sulfamethoxazole in WWTP discharge before and after PFA disinfection, as characterized by suspect screening in UPLC-IMS-QTOF. The intensity of each organic micropollutant was normalized with the intensity of propylparaben-d4 as an internal standard. error bars represent the standard deviation of intensity values over triplicates.

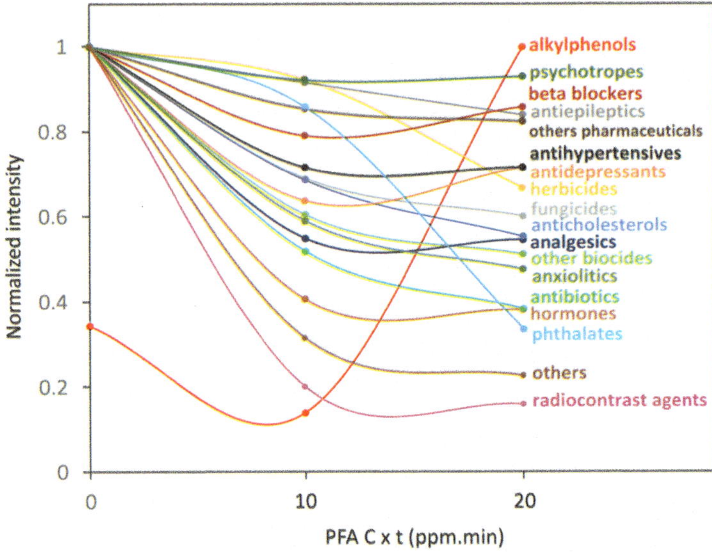

Figure 25 Evolution of categories of organic micropollutants during PFA disinfection (10 and 20 ppm.min) of treated water discharge from the seine amont valenton WWTP (SEV-TW6, Nov 6, 2018), as characterized by suspect screening in UPLC-IMS-QTOF. The intensity of each organic micropollutant was first normalized with the intensity of propylparaben-d4 as an internal standard, then the normalized intensity was calculated as the average per category and over triplicates, divided by the maximum intensity value in the category.

SECTION 2 – CHAPTER 1

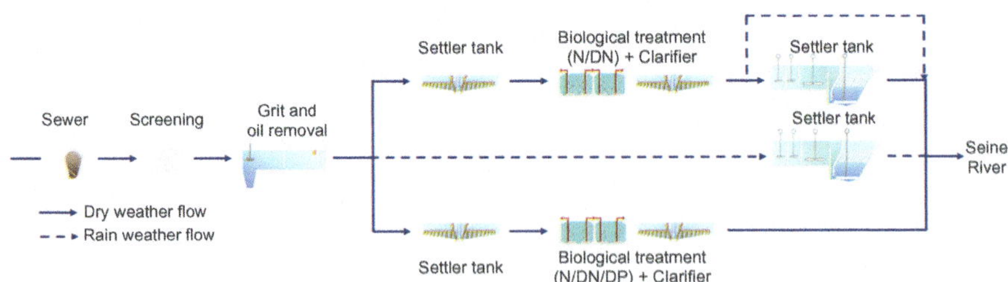

Figure 26 Seine Valenton WWTP treatment flow chart.

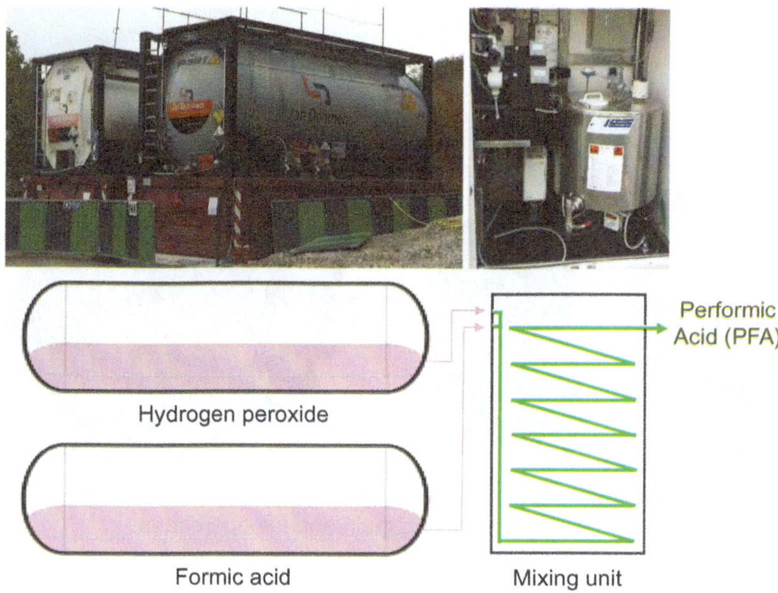

Figure 27 Kemira desinfix® (performic acid, PFA) production process.

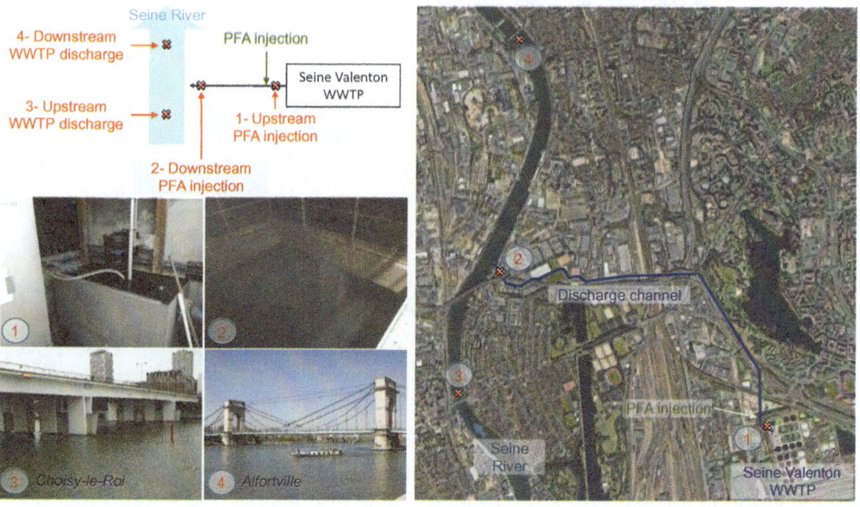

Figure 28 Sampling points on the Seine Valenton WWTP discharge channel (upstream and downstream of PFA injection) and sampling sites on the seine river (upstream and downstream of WWTP discharge).

SECTION 2 – CHAPTER 2

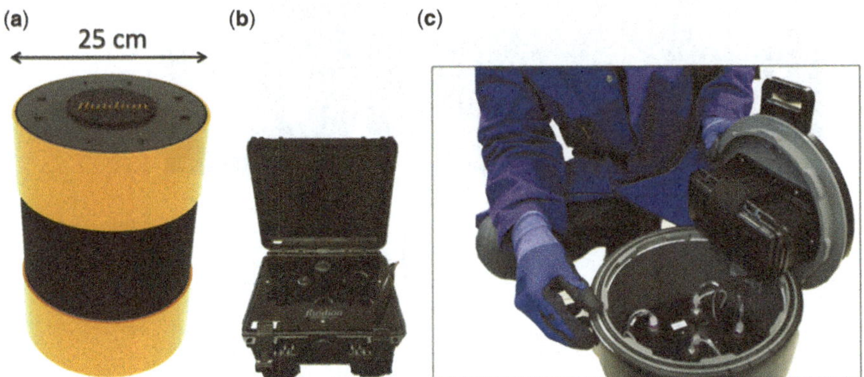

Figure 29 (a) *In situ* ALERT system; (b) ALERT LAB portable/bench-top instrument; (c) view of the ALERT system during field maintenance operations.

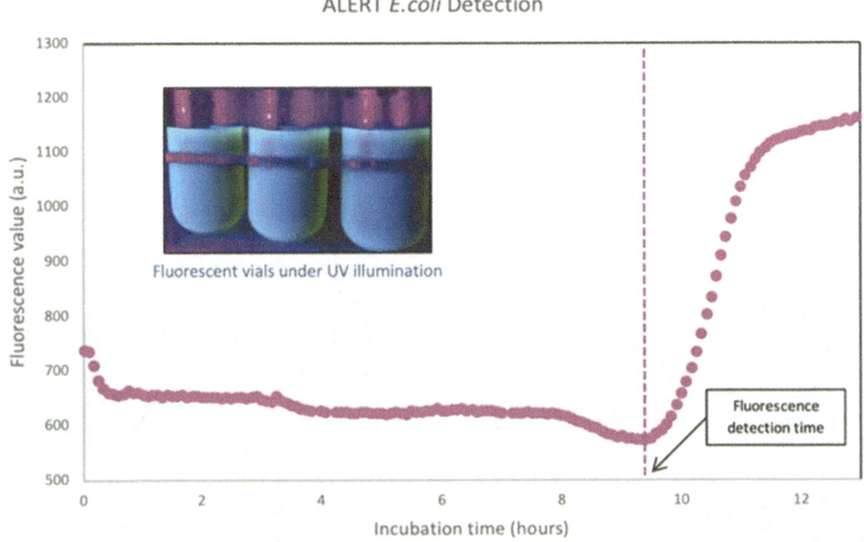

Figure 30 Typical fluorescence signal obtained from the ALERT device: an initial signal plateau is followed by a sharp increase in fluorescence starting a few hours into the measurement. The fluorescence detection time is automatically interpreted by the cloud-based server and used to provide an *E. coli* quantification. inset: photo of sample vials illuminated by UV light, after incubation.

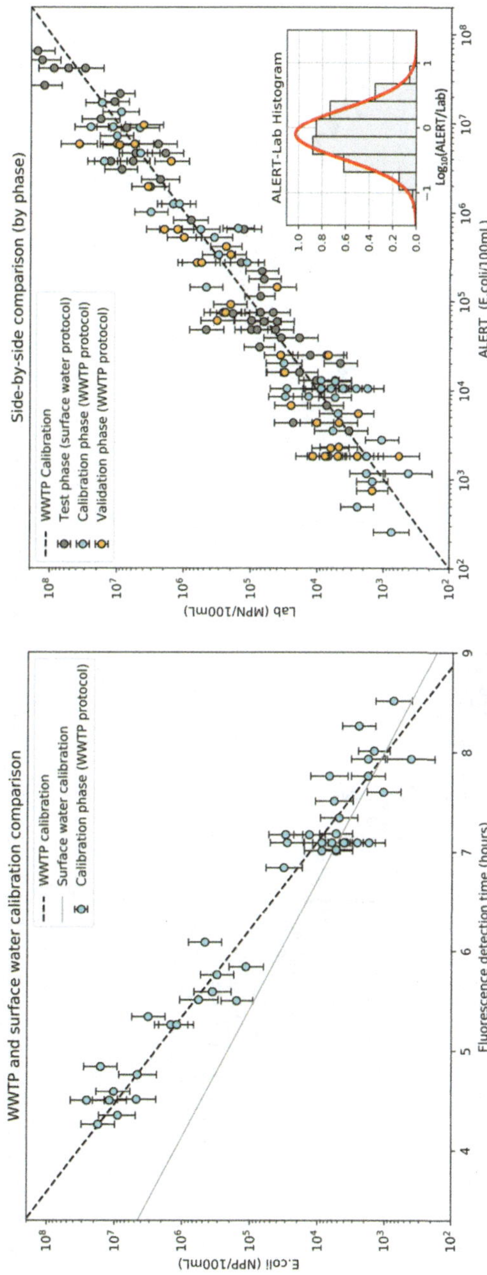

Figure 31 Left: Side-by-side results obtained during the calibration phase, along with the previous surface water calibration (solid gray line) and the newly developed WWTP calibration (dashed line). Right: Side-by-side results grouped by phase (test, calibration, validation). Inset: The alert/Lab offset histogram.

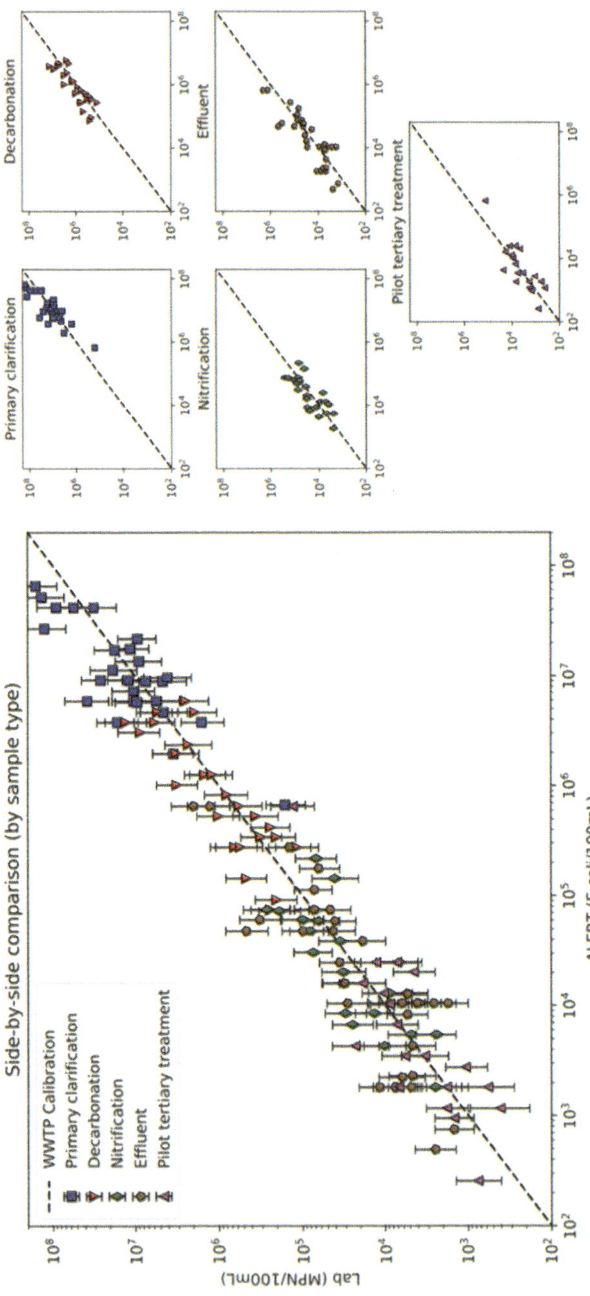

Figure 32 Side-by-side data grouped by sample origin. All WWTP protocol data are displayed (calibration and validation phases). The right-hand panels focus on samples from individual treatment stages, as indicated.

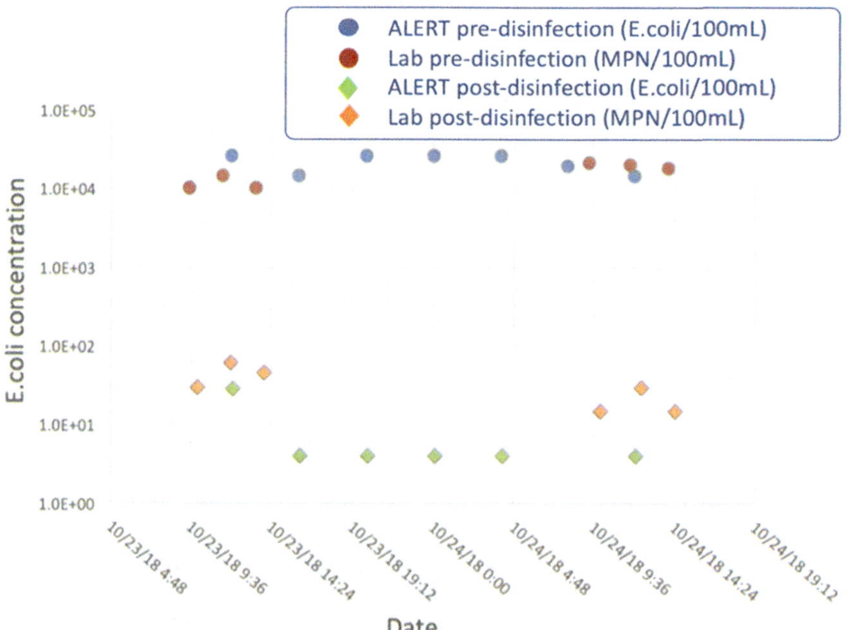

Figure 33 24 hours of ALERT pre-disinfection (in blue) and post-disinfection (green) measurements at the valenton WWTP disinfection pilot. comparative laboratory results at the beginning and end of the 24-hour period are also shown in red and orange, respectively.

SECTION 2 – CHAPTER 3

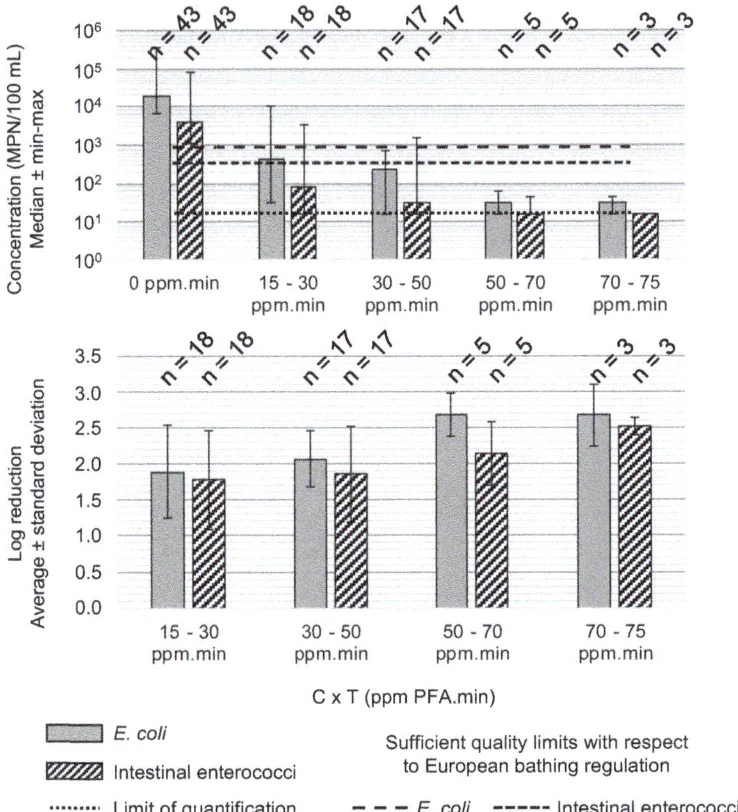

Figure 34 Effectiveness of fecal bacteria disinfection by PFA applied in-line at the industrial scale to the SEV WWTP discharge.

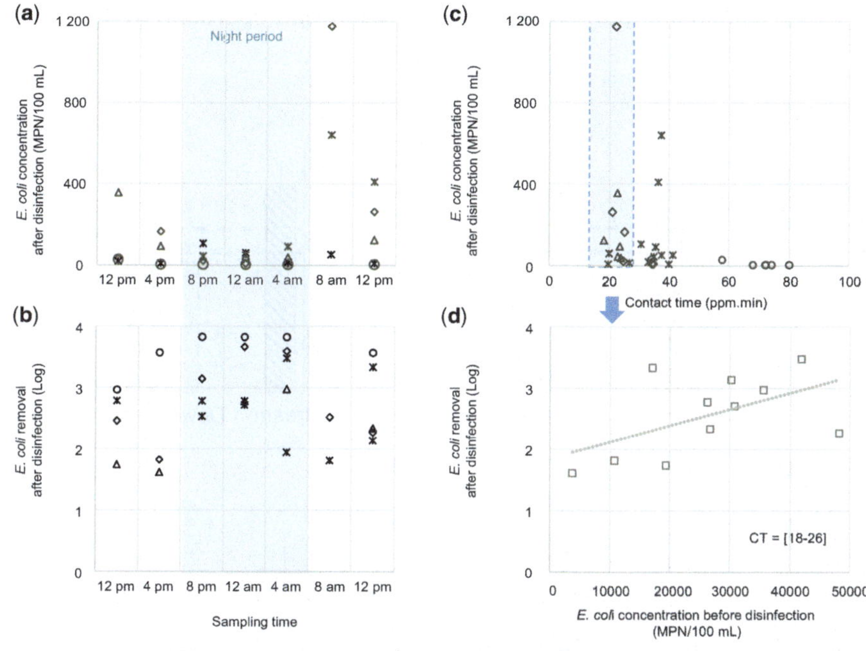

Figure 35 Variations in *E. coli* values after disinfection at different PFA concentrations (0.8, 1, 1.2 and 2 ppm): (a) concentrations (MPN/100 mL) at different times of day, and (b) the associated removal rate (log); (c) concentrations (MPN/100 mL) for different contact time values (ppm.min), and (d) the removal rate for a fixed contact time of 18–26 ppm.min.

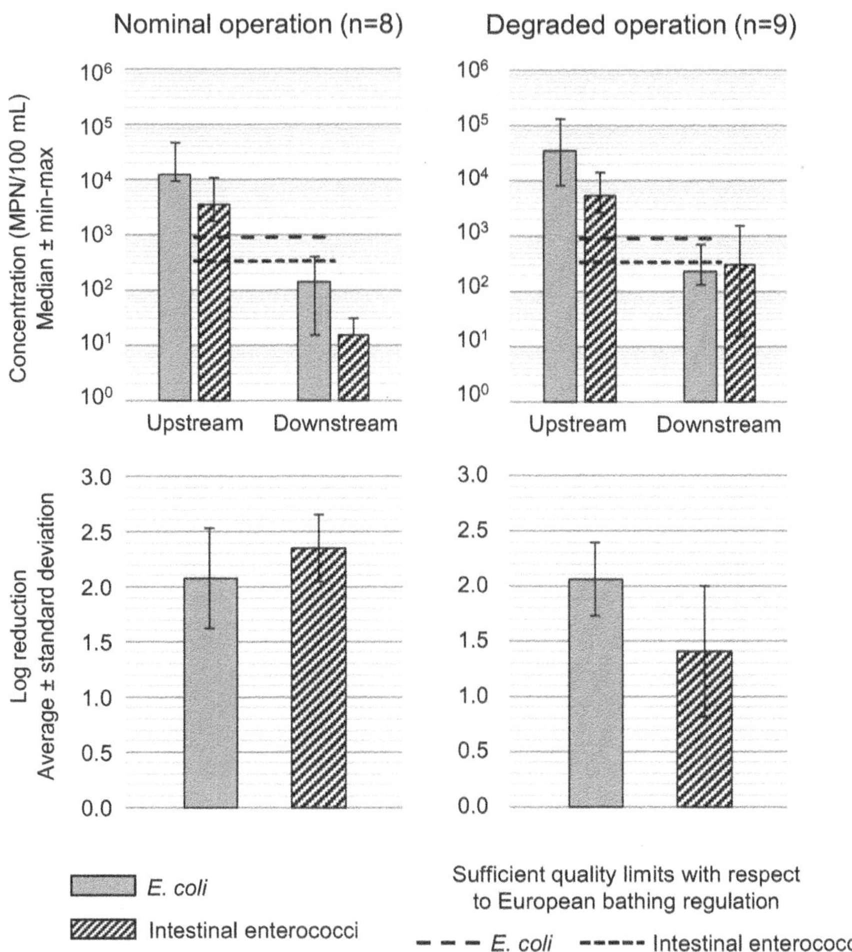

Figure 36 Fecal bacteria disinfection effectiveness with PFA applied at the industrial scale at SEV WWTP under nominal or degraded operations.

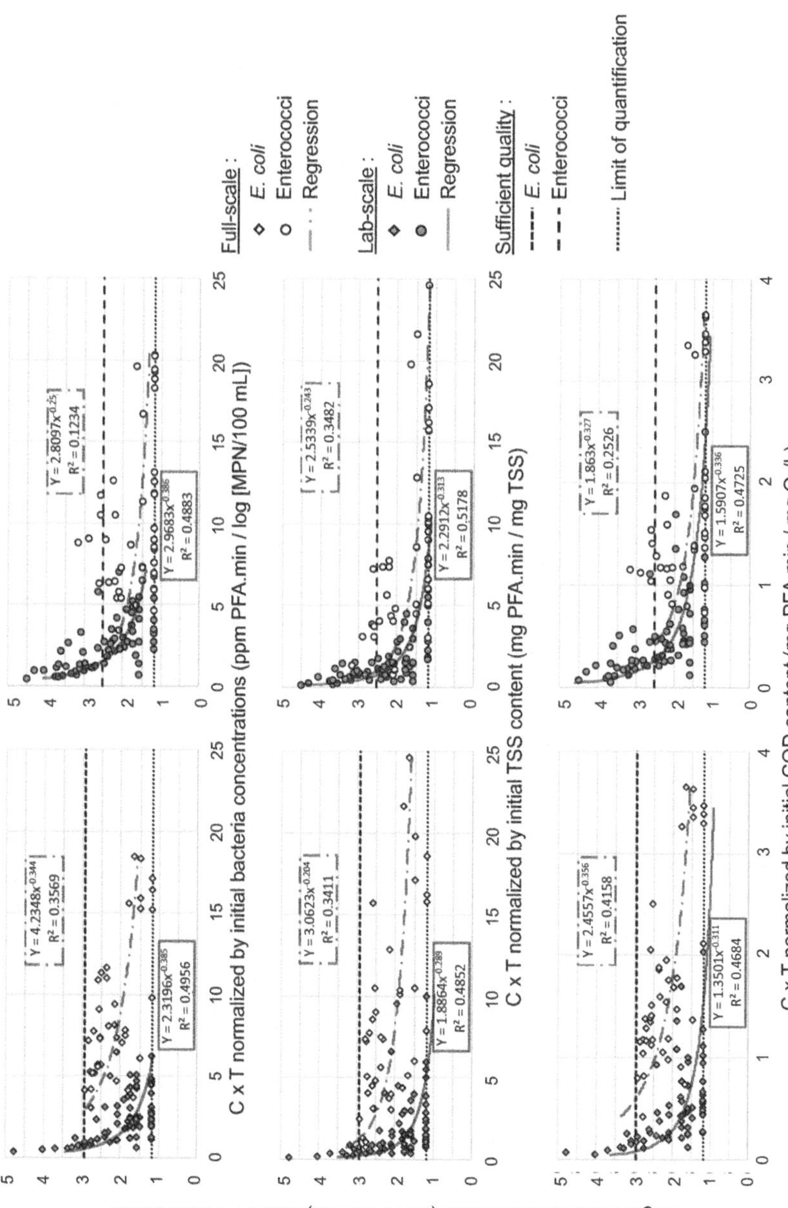

Figure 37 Evaluation of PFA disinfection effectiveness at the industrial scale by normalizing the applied PFA dose to the initial bacterial concentrations, TSS and COD.

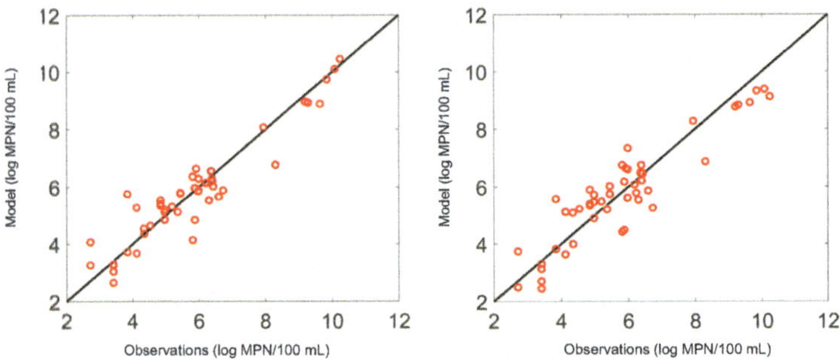

Figure 38 Parity plots of the model estimation results for optimal models F3 (left) and S2 (right). observations and predictions are natural log-values.

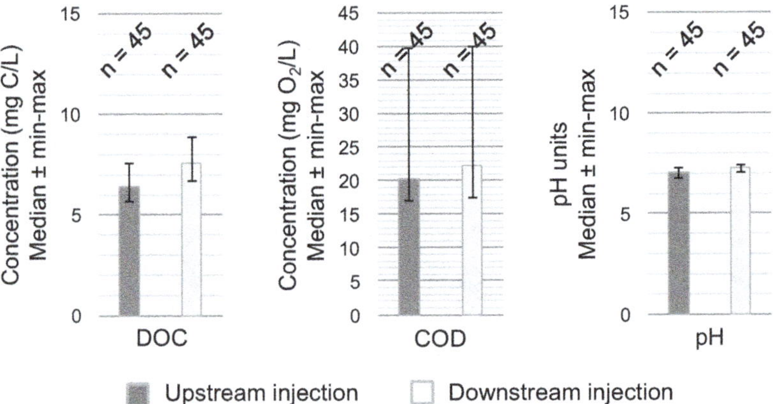

Figure 39 Evolution of DOC, COD and pH of SEV WWTP discharge with PFA disinfection.

SECTION 3 – CHAPTER 1

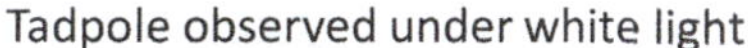

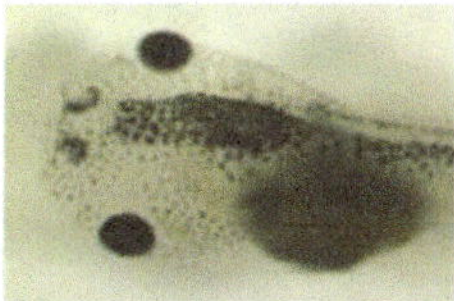

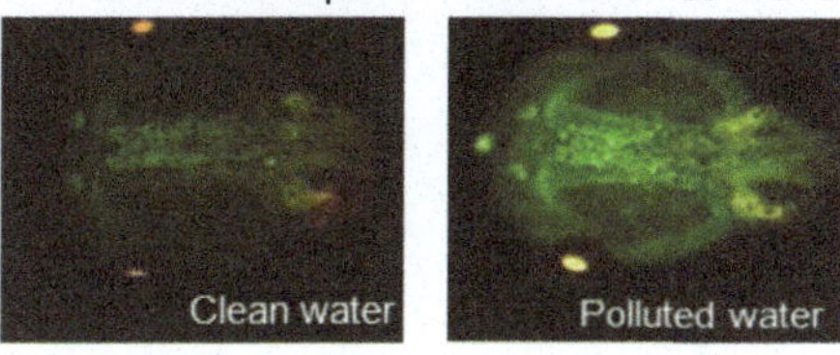

Figure 40 Photographs of tadpoles used as bio-indicators.

Figure 41 Results obtained for the endocrine disruption tests.

SECTION 3 – CHAPTER 2

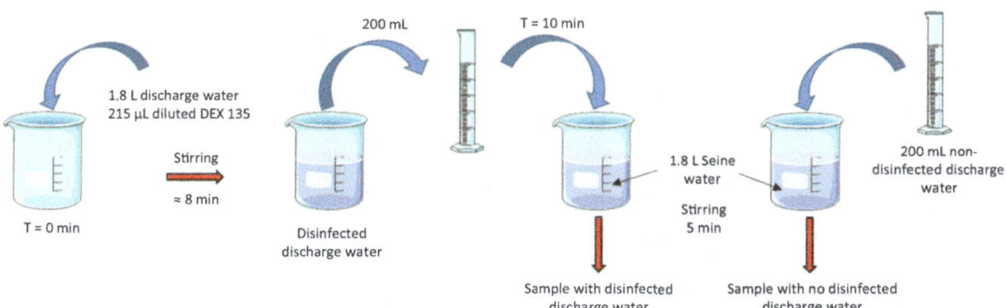

Figure 42 Schematic diagram of the experimental protocol for test sample preparation.

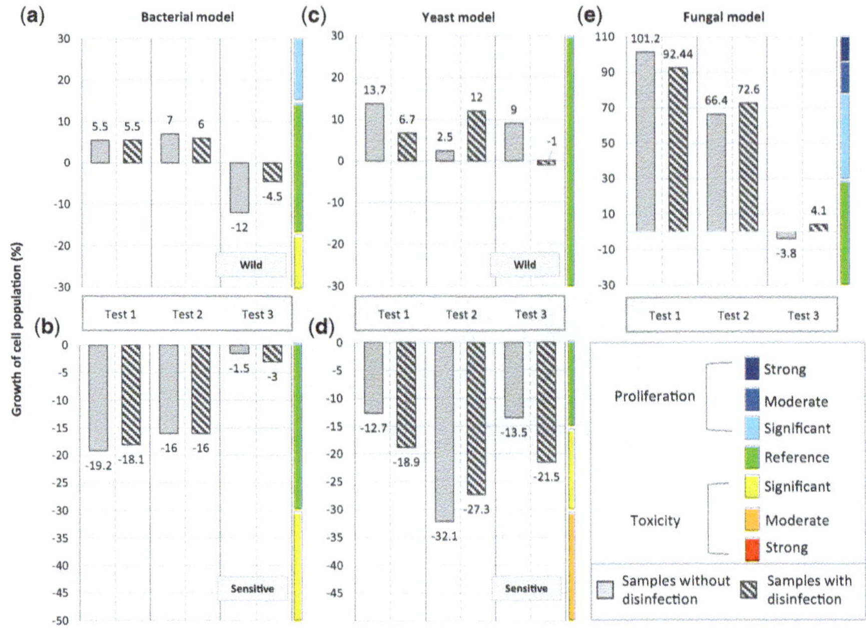

Figure 43 Laboratory-scale results on the general toxicity of Seine water samples, supplemented with disinfected and non-disinfected discharge water.

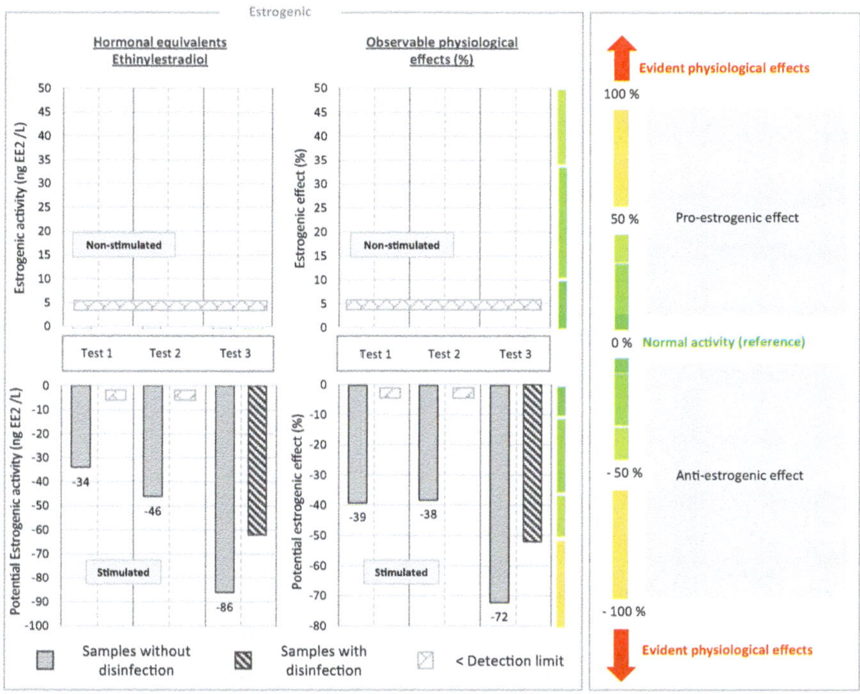

Figure 44 Laboratory-scale results on estrogenic disruption in samples of Seine River water supplemented with disinfected and non-disinfected discharge water.

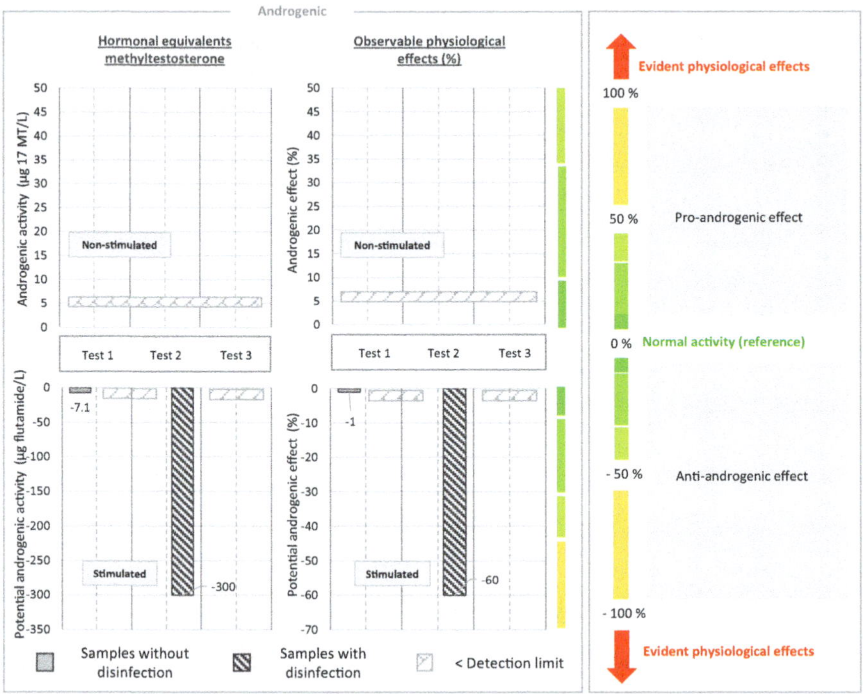

Figure 45 Laboratory-scale results on androgenic disruption in samples of Seine water supplemented with disinfected and non-disinfected discharge water.

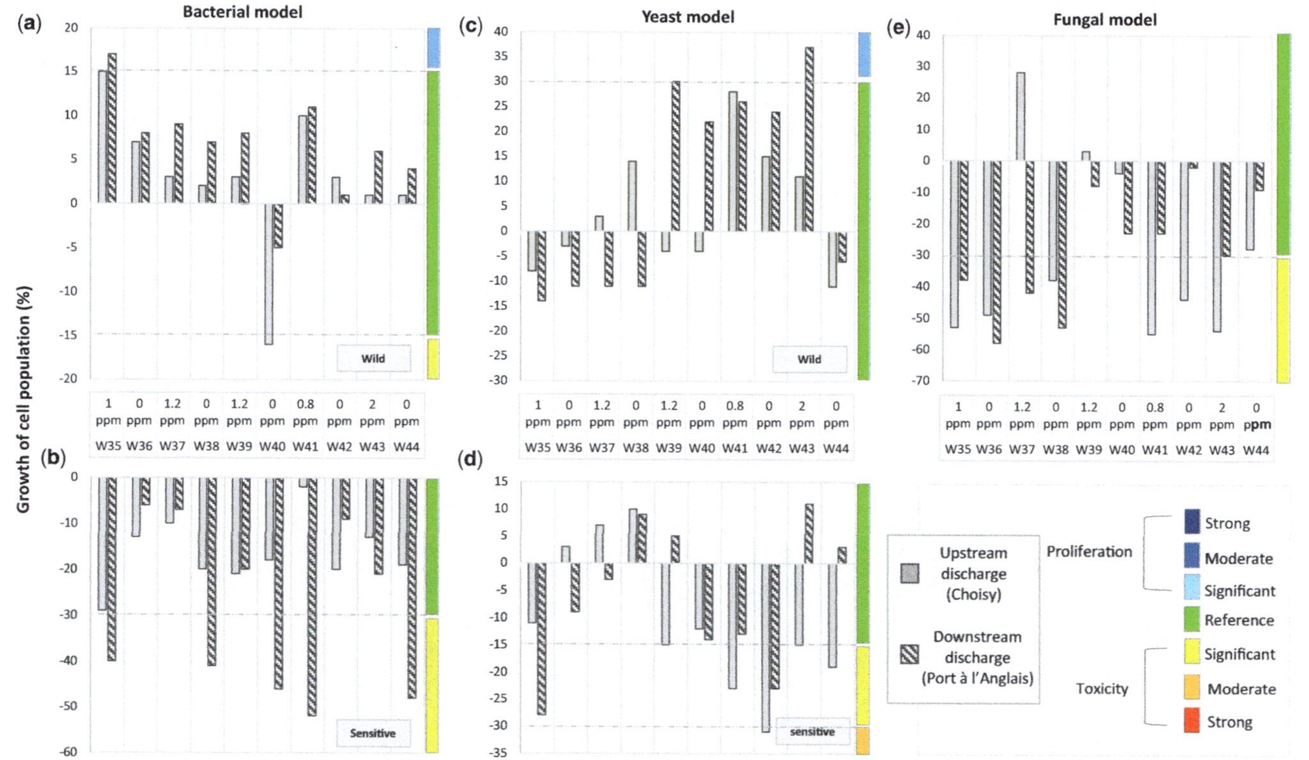

Figure 46 General toxicity assessed in the Seine River upstream and downstream of the SEV outfall during industrial-scale testing.

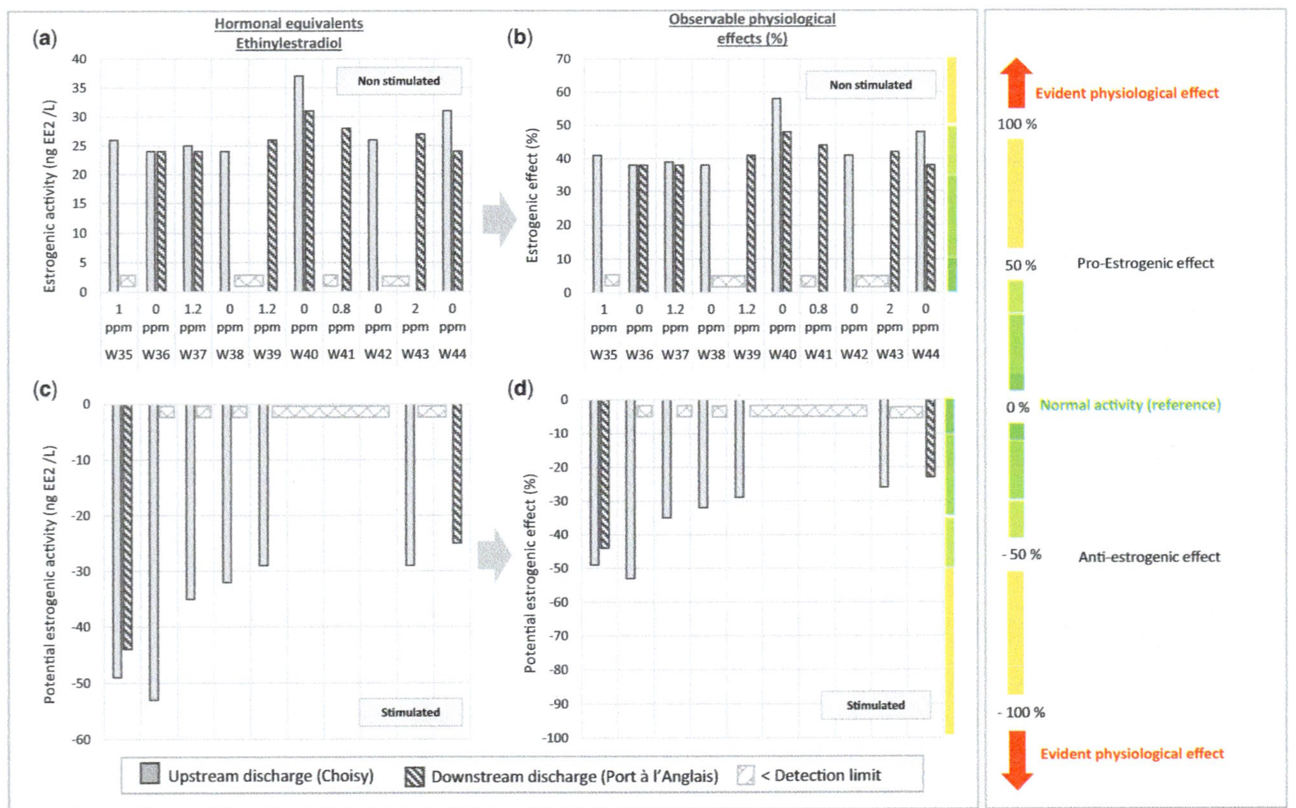

Figure 47 Estrogenic effects assessed in the seine river upstream and downstream of the SEV outfall during industrial-scale testing.

SECTION 4 – CHAPTER 1

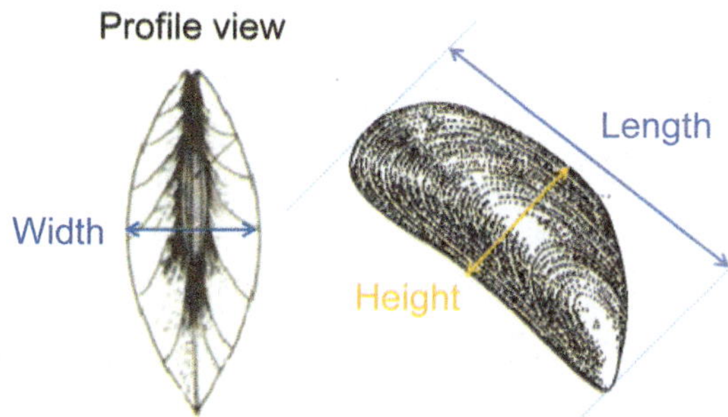

Figure 48 Main morphometric measurements in bivalves: length = anterior-posterior length, height = dorsoventral distance, and width = distance between the right and left valve.

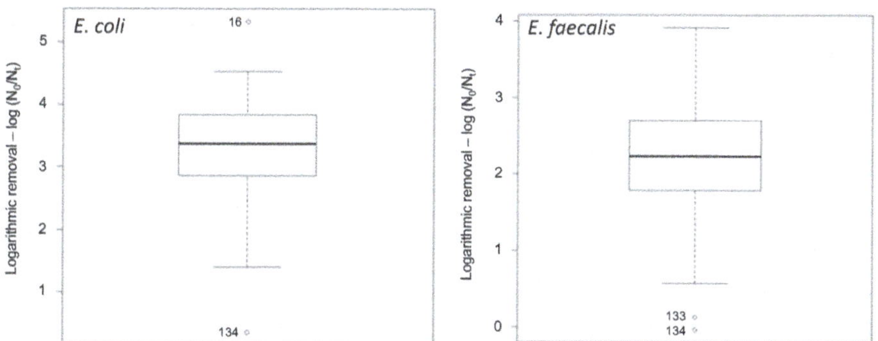

Figure 49 Boxplot of the removal (expressed in u.log) of *E. coli* (left) and *E. faecalis* (right) by means of tertiary treatment by PFA (kemConnect™) ($N = 145$ samples).

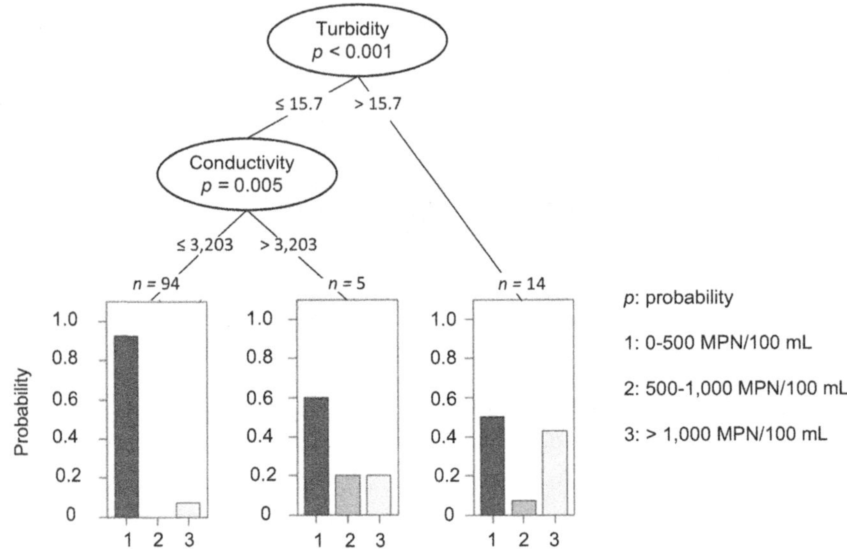

Figure 50 Conditional tree obtained upon analysis of removal results for *E. coli* after tertiary treatment according to various parameters (e.g., turbidity, conductivity, temperature).

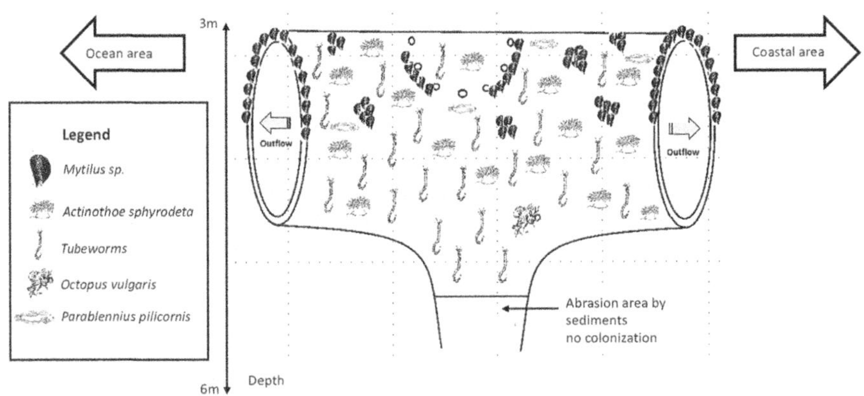

Figure 51 Distribution of species observed on the WWTP outfall.

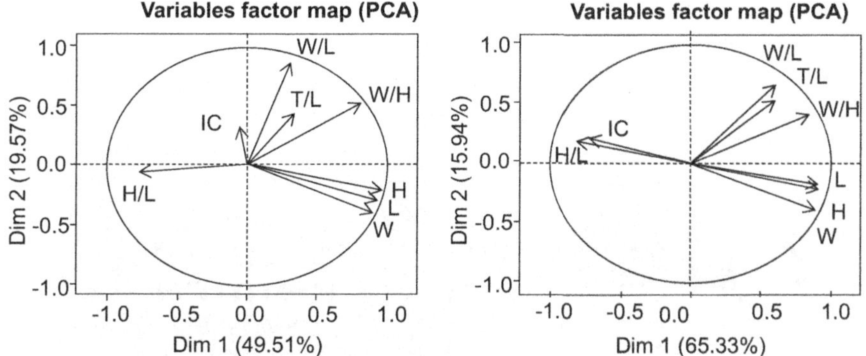

L = Length, H = Height, W = Width, T = Thickness, H/L = Elongation index, W/L = Compactness index, W/H = Convexity index, T/L = Thickness index

Figure 52 Comparisons of the morphometric characteristics of individuals collected from the outfall in 2015 (left) and 2017 (right).

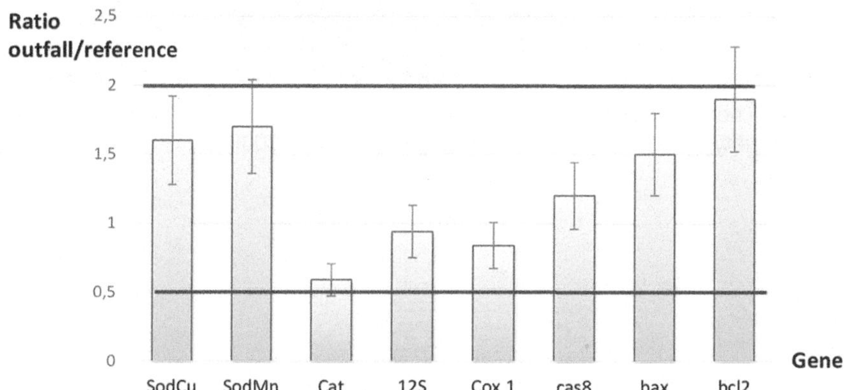

Figure 53 Ratio of the expression of different genes between mussels collected from the outfall and a reference site on the same date (sodCu and sodMn = oxidant stress genes; rad51 = DNA repair gene, Cox1 = metabolism gene, cas8, bax and bcl 2 = apoptosis genes, red lines = normal variability for a population).

SECTION 4 – CHAPTER 2

Figure 54 The first PFA *hyproform* production system (kemira Oyj) used in full-scale studies: prototype (a, b – 2005) and pilot system (c, d – 2006).

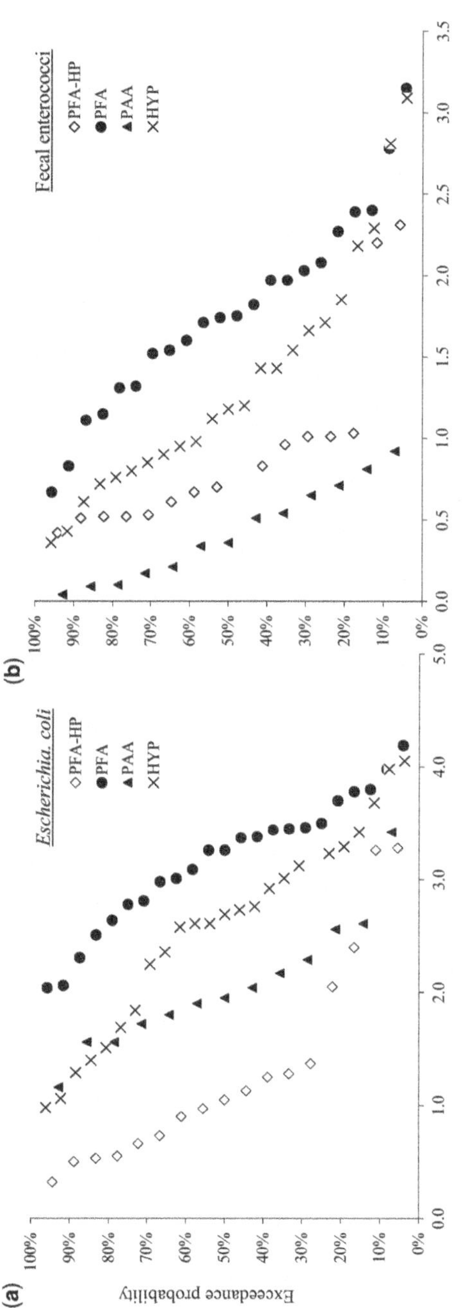

Figure 55 Comparisons of the probability of disinfectants to achieve and exceed the log bacterial reductions, at a retention time ranging from 10 to 20 minutes in the disinfection reactor. Log(N_0/N) = log reduction; N: bacterial concentration; N_0: initial bacterial concentration; PFA = desinfix performic acid solution; PFA-HP = hyproform performic acid solution; PFA-HP = hyproform performic acid solution.

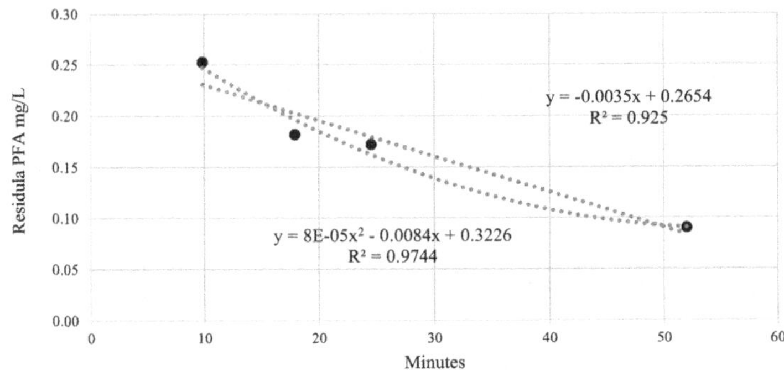

Figure 56 Average of PFA residuals vs. contact time in the disinfection reactor of eraclea, jesolo, San donà and caorle, respectively; measurements performed in 2016 and 2017.

List of Tables

SECTION 1 – CHAPTER 1

doi: 10.2166/9781789062106_0201

Table 1 Impact on SSR and F-specific bacteriophages of PFA disinfection applied to SEV WWTP discharge (contact time: 10 min).

Date	SSR Spores				F-Specific Bacteriophages			
	Concentration (CFU/100 mL)			Log Removal	Concentration (CFU/100 mL)			Log Removal
	SEV WWTP Discharge	8–12 ppm.min of PFA	20 ppm. min of PFA	8–20 ppm. min of PFA	SEV WWTP Discharge	8–12 ppm.min of PFA	20 ppm. min of PFA	8–20 ppm. min of PFA
Sept 11, 2018	80	7	2	1.06–1.60	33		30	No
Sept 18, 2019*	5200	89	<1	1.77–>3.72	33		<30	>0.04
Jan 30, 2019*	6400	4700		0.13	<30	<30		?
Feb 6, 2019	6400	6500		No	<30	<30		?
Feb 13, 2019	2170	5000		No	<30	<30		?
May 5, 19	1800	1500	675	0.08–0.43				
May 9, 2019	5000	6300	5900	No				
Median	5000	4700	339		<30	<30	<30	

*Days with degraded SEV WWTP operations.

SECTION 1 – CHAPTER 3

Table 2 Excitation and emission wavelengths of DOM fluorophores and their interpretation.

Fluorophores (Parlanti *et al.*, 2000)	Excitation Wavelength (nm)	Emission Wavelength (nm)	Type of Compound
α'	230–260	380–480	Humic-like substances + more recent material
α	330–370	420–480	Humic-like substances
β	310–320	380–420	Recent material + biological components
γ	270–280	300–320	Tyrosine-like
δ	270–280	320–380	Tryptophan-like

Table 3 PFA half-life measured in both SEV WWTP discharge and seine river water in choisy samples spiked with 2.5 ppm of PFA.

Type of Water	Sampling Date	Temperature (°C)	Filtration at 0.45 μm	$t_{1/2}$ (min)
SEV WWTP discharge	Sept 11, 2018	12	Yes	33
		20		17
		25		13
	Oct 16, 2018	20	Yes	29
			No	29
	Nov 6, 2018		No	33
	Average ± standard deviation	20	–	26 ± 9
Seine River water in Choisy	Sept 11, 2018	12	Yes	87
		20		53
		25		29

SECTION 1 – CHAPTER 4

Table 4 N-nitrosamines monitored in this study and their optimized analytical parameters for detection by gas chromatography coupled with tandem mass spectrometry (GC-MS/MS).

Compound[a]	RT (min)[b]	Parent ion	Product ions (collision energy)[c]
N-nitrosodimethylamine-d6 (NDMA-d6)	6.68	80	50 (10), 46 (20)
N-nitrosodimethylamine (NDMA)	6.70	74	44 (5), 42 (15)
Toluene-d8	6.94	98	70 (15), 98 (10)
N-nitrosomethylethylamine (NMEA)	7.50	88	71 (5), 43 (10)
N-nitrosodiethylamine (NDEA)	8.15	102	85 (5), 44 (10)
N-nitrosodipropylamine-d14 (NDPA-d14)	9.68	144	126 (5), 50 (10)
N-nitrosodipropylamine (NDPA)	9.76	130	113 (5), 43 (10)
N-nitrosopyrrolidine (NPYR)	9.76	100	55 (5), 43 (10)
N-nitrosomorpholine (NMOR)	9.79	116	86 (5), 56 (10)
N-nitrosopiperidine (NPIP)	10.23	114	84 (5), 42 (15)
N-nitrosodibutylamine (NDBA)	11.57	158	99 (10), 141 (5)
N-nitrosodiphenylamine (NDPhA)	15.25	168	167 (20), 166 (25)

[a]NDMA-d6 was used as the internal standard for NDMA quantification; NDPA-d14 was used as the internal standard for other N-nitrosamine quantification; Toluene-d8 served as the injection internal standard used to monitor analytical stability and calculate the recovery rate.
[b]RT = Retention time.
[c]The first product ion was used for quantification.

Table 5 Formation of AOX and halogenated DBPs from the PFA disinfection of WWTP water at the laboratory scale (various concentrations of PFA, 10-min contact time) or during full-scale disinfection (0.8–2.5 ppm PFA).

Date	Sample[a]	C.t (ppm. min)	AOX (µg/L)	DBPs[b] (µg/L)
Laboratory-scale disinfection trials				
Sept 11, 2018	SEV-TW	–	39.4	n.a[c]
	SEV-TW + 2 ppm PFA	20	48.7	n.a

(_Continued_)

Table 5 Formation of AOX and halogenated DBPs from the PFA disinfection of WWTP water at the laboratory scale (various concentrations of PFA, 10-min contact time) or during full-scale disinfection (0.8–2.5 ppm PFA) (*Continued*).

Date	Sample[a]	C.t (ppm. min)	AOX (µg/L)	DBPs[b] (µg/L)
Sept 18, 2018	SEV-TW	–	32.0	n.d[d]
	SEV-TW + 2 ppm PFA	20	43.4	n.d
Nov 6, 2018	SEV-TW	–	45.0	n.d
	SEV-TW + 1 ppm PFA	10	40.6	n.d
	SEV-TW + 2 ppm PFA	20	48.9	n.d
	SEV-TW + 30 ppm PFA	300	49.2	DBCM (0.02), BCAN (0.14)
	SEV-TW + 100 ppm PFA	1000	338.7	DCBM (2.35), DBCM (1.39), TBM (0.03), DCAN (8.25), BCAN (1.96), DBAN (0.55), TCAN (0.73), DCP (3.41), TCNM (1.83)
	SEV-TW + 2 mg-N/L of NO_2^- + 1 ppm PFA	10	32.6	BCAN (0.15), DBAN (0.10)
Dec 11, 2018	SEC-RW	–	41.7	DBCM (0.34), TBM (0.23)
	SEC-RW + 10 ppm PFA	100	39.5	DBCM (0.43), TBM (0.36), DBAN (0.16)
	SEC-SW	–	42.9	DBCM (0.34), TBM (0.24), DBAN (0.14)
	SEC-SW + 10 ppm PFA	100	39.9	DBCM (0.29), TBM (0.26), DBAN (0.12)
Full-scale disinfection trials				
Sept 26, 2018	SEV-TW	–	31.2	n.d
	SEV-DW (1.2 ppm PFA)	32.1	44.6	n.d
Oct 10, 2018	SEV-TW	–	43.9	n.d
	SEV-DW (0.8 ppm PFA)	28.3	49.9	n.d

Table 5 Formation of AOX and halogenated DBPs from the PFA disinfection of WWTP water at the laboratory scale (various concentrations of PFA, 10-min contact time) or during full-scale disinfection (0.8–2.5 ppm PFA) (*Continued*).

Date	Sample[a]	C.t (ppm. min)	AOX (µg/L)	DBPs[b] (µg/L)
Oct 24, 2018	SEV-TW	–	47.2	BCAN (0.34), DBAN (0.69)
	SEV-DW (2.5 ppm PFA)	74.2	54.8	DBCM (0.07), BCAN (0.43), DBAN (1.1)

[a]SEV: Seine Amont Valenton WWTP; SEC: Seine Centre WWTP; TW: treated water; DW: disinfected water at the full-scale; RW: pretreated raw water; SW: settled water; PFA: performic acid.

[b]DCBM: dichlorobromomethane; DBCM: dibromochloromethane; TBM: bromoform; DCAN: dichloroacetonitrile; BCAN: bromochloroacetonitrile; DBAN: dibromoacetonitrile; TCAN: trichloroacetonitrile; DCP: dichloropropanone; TCNM: chloropicrin.

[c]n.a: not analyzed.

[d]n.d: not detected.

Table 6 Concentration of bromide ions in WWTP discharge before and after disinfection with PFA (0.8–2.5 ppm PFA).

Date	C × t (ppm.min)	Bromide ions (µg/L)	
		Before PFA Injection	After PFA Injection
Aug 28, 2018	31.2	100	90
Sept 11, 2018	31.6	120	50
Sept 25, 2018	36.4	250	160
Oct 9, 2018	22.1	170	170
Oct 23, 2018	73.8	170	170

Table 7 N-nitrosamine concentrations in WWTP water disinfected by PFA at the laboratory scale (various concentrations of PFA, 10-min contact time) or during full-scale disinfection (0.8–2.5 ppm PFA).

Date	Sample[a]	$C \times t$ (ppm.min)	N-nitrosamines[b] (ng/L)					
			NDMA	NDEA	NDPA	NMOR	NPIP	NDBA
Laboratory-scale disinfection trials								
Sept 11, 2018	SEV-TW	–	30	3.5	2.6	11	1.1	4.3
	SEV-TW + 2 ppm PFA	20	25	1.0	0.2	9.8	0.7	1.7
Sept 18, 2018	SEV-TW	–	31	3.0	1.5	13	2.1	7.7
	SEV-TW + 2 ppm PFA	20	22	2.2	0.3	12	1.2	2.8
Nov 6, 2018	SEV-TW	–	21	1.5	1.5	10	1.1	5.6
	SEV-TW + 1 ppm PFA	10	21	2.1	0.8	14	1.2	2.8
	SEV-TW + 2 ppm PFA	20	24	3.9	0.8	8.8	1.9	2.7
	SEV-TW + 30 ppm PFA	300	24	2.0	0.2	9.1	1.4	2.8
	SEV-TW + 100 ppm PFA	1000	19	2.8	1.9	11	2.1	8.8
	SEV-TW + 2 mg-N/L of NO_2^- +1 ppm PFA	10	30	4.9	1.0	162	1.2	3.4
Dec 11, 2018	SEC-RW	–	27	19	0.3	7.2	4.5	2.6
	SEC-RW + 10 ppm PFA	100	33	23	0.2	6.4	8.6	2.9
	SEC-SW	–	30	17	0.4	7.4	6.2	1.8
	SEC-SW + 10 ppm PFA	100	28	18	0.6	7.8	12.5	1.9
Full-scale disinfection trials								
Sept 26, 2018	SEV-TW	–	32	14	0.6	11	1.6	6.7
	SEV-DW (1.2 ppm PFA)	32.1	31	11	1.5	14	2.4	19
Oct 10, 2018	SEV-TW	–	31	4.9	1.5	18	3.2	5.2

(Continued)

Table 7 N-nitrosamine concentrations in WWTP water disinfected by PFA at the laboratory scale (various concentrations of PFA, 10-min contact time) or during full-scale disinfection (0.8–2.5 ppm PFA) (*Continued*).

Date	Sample[a]	C × t (ppm.min)	N-nitrosamines[b] (ng/L)					
			NDMA	NDEA	NDPA	NMOR	NPIP	NDBA
	SEV-DW (0.8 ppm PFA)	28.3	31	6.5	0.7	19	3.1	11
Oct 24, 2018	SEV-TW	–	33	3.9	1.8	15	3.2	6.6
	SEV-DW (2.5 ppm PFA)	74.2	33	5.3	0.5	13	1.7	6.7

[a]SEV: Seine Amont Valenton WWTP; SEC: Seine Centre WWTP; TW: treated water; DW: disinfected water at full-scale; RW: pretreated raw water; SW: settled water; PFA: performic acid.
[b]NDMA: N-nitrosodimethylamine; NDEA: N-nitrosodiethylamine; NDPA: N-nitrosodipropylamine; NMOR: N-nitrosomorpholine; NPIP: N-nitrosopiperidine; NDBA: N-nitrosodibutylamine.

Table 8 Bulk characterization of marker fingerprints obtained from WWTP discharge samples before (SEV-TW) and after PFA disinfection (SEV-DW and SEV-TW + 2 ppm PFA) analyzed by UPLC-IMS-QTOF in positive mode.

	Before PFA SEV-TW[a]		After PFA SEV-DW[b] and SEV-TW + 2 ppm PFA	
	All Markers	Unique Markers	All Markers	Unique Markers
Average RT[c] (min)	8.96	10.26	9.09	11.03
Average m/z ratio	434.4252	505.4057	405.3870	441.2656
Total number of markers	11,608	1799	11,634	2119
Number of markers with intensity >10,000	7369	1792	6222	1303
Total intensity of markers	4.8×10^8	4.5×10^7	4.1×10^8	4.1×10^7
Total intensity of markers with intensity >10,000	4.6×10^8	6.0×10^7	3.8×10^8	5.5×10^7

All values are averaged from six triplicate samples (18 fingerprints). The parameters were calculated from all markers detected in each of the 18 fingerprints, as well as from markers exclusively detected before or after PFA disinfection ('unique markers').
[a]SEV-TW: Seine Amont Valenton WWTP treated water.
[b]SEV-DW: Seine Amont Valenton WWTP disinfected water at the full scale.
[c]RT: retention time.

SECTION 2 – CHAPTER 1

Table 9 PFA disinfection conditions applied during the 10-week, full-scale trial in 2018.

Weeks	35	36	37	38	39	40	41	42	43	44
Targeted PFA injection dose (ppm)	1	0	1.2	0	1.2	0	0.8	0	2	0
Actual PFA injection dose (ppm)	0.8–1.0	0	0.8–1.2	0	1.2	0	0.8	0	2.1–2.5	0
Hydraulic retention time (min)	22–31	19–56	25–39	25–36	27–38	20–54	22–35	24–37	28–34	23–40
$C \times t$ (ppm.min)	17–31	0	16–39	0	32–45	0	18–28	0	60–74	0

Table 10 Sampling conducted and parameters monitored during the 10-week, full-scale trial.

Compound	SEV WWTP Discharge						Seine River		
	Upstream Injection All Weeks			Downstream Injection Weeks with Injection			Upstream and Downstream of WWTP Discharge All Weeks		
	Tu	We	Th	Tu	We	Th	Tu	We	Th
E. coli	3	3	3	3	3	3		1	
Intestinal enterococci	3	3	3	3	3	3		1	
Spores of anaerobic sulfite-reducing microorganisms	1			1					
F-RNA specific bacteriophages	1			1					
TSS	3	3	3	3	3	3		1	
DOC	3	3	3	3	3	3		1	
COD	3	3	3	3	3	3		1	
BOD	3	3	3	3	3	3		1	
TKN	3	3	3	3	3	3		1	
$N-NH_4$	3	3	3	3	3	3		1	
$N-NO_2$	3	3	3	3	3	3		1	
$N-NO_3$	3	3	3	3	3	3		1	
Total phosphorus	3	3	3	3	3	3		1	
$P-PO_4$	3	3	3	3	3	3		1	
pH	3	3	3	3	3	3		1	

(Continued)

Table 10 Sampling conducted and parameters monitored during the 10-week, full-scale trial *(Continued)*.

Compound	SEV WWTP Discharge						Seine River		
	Upstream Injection All Weeks			Downstream Injection Weeks with Injection			Upstream and Downstream of WWTP Discharge All Weeks		
	Tu	We	Th	Tu	We	Th	Tu	We	Th
Conductivity	3	3	3	3	3	3	1		
Turbidity	3	3	3	3	3	3	1		
Halogenated organic hydrocarbons (AOX)	1			1	0	0			
Bromate	1			1					
Bromide	1			1					
Color								1	
Chloride								1	
Sulfate								1	

Table 11 Analytical methods used for the determination of the monitored parameters.

Analytical Parameter	Applicable Standards	Limit of Quantification	Analytical Uncertainty
Ammonium (NH_4^+)	NF EN ISO 11732 (August 2005)	0.3 mg N/L	0.3–1.2 = 40% >1.2 = 10%
	NF EN ISO 11732 (August 2005)	0.01 mg NH_4/L	0.010–0.025 = 50% >0.025 = 20%
Halogenated organic hydrocarbons (AOX)	NF EN ISO 9562 (March 2005)	10 µg/L Cl	>10 = 35%
F-RNA specific bacteriophage (F-RNA)	ISO 10705−3/NF EN ISO 10705−1	30 PFU/100 mL	Undetermined
Bromate	ISO 10304−1	5 µg/L	Undetermined
Bromide	Internal method	50 µg/L	Undetermined
Total Kjeldahl nitrogen (TKN)	NF EN 25663 (January 1994) or SIAAP internal method	0.5 mg N/L	0.5–2 = 60% >2 = 15% or 0.5–1.5 = 60% >1.5 = 20%
Dissolved organic carbon (DOC)	NF EN 1484 (July 1997)	0.3 mg C/L (chemical method) or 3 mg/L (thermal method)	0.3–0.8 = 40% >0.8 = 15% or 3–12 = 60% >12 = 15%
Chloride	NF ISO 15923−1 (January 2014)	5 mg Cl/L	5–20 = 40% >20 = 10%
	NF EN ISO 10304−1 (July 2009)	1 mg Cl/L	1–1.5 = 15% >1.5 = 10%
Conductivity	NF EN 27888 (January 1994)	30 mS/m	>30 = 10%
Water color	NF EN ISO 7887	0 mg Pt/L	0–15 = ± 4.5 mg Pt/L >15 = 30%

(Continued)

Table 11 Analytical methods used for the determination of the monitored parameters (*Continued*).

Analytical Parameter	Applicable Standards	Limit of Quantification	Analytical Uncertainty
Biochemical oxygen demand (BOD$_5$)	NF EN 1899−1 (May 1998) or NF EN 1899−2 (May 1998)	3 mgO$_2$/L or 0.5 mgO$_2$/L	3–4 = 40% >4 = 30% or 0.5–6 = 30%
Chemical oxygen demand (COD)	ISO 15705 (November 2002)	4 mgO$_2$/L	4–6.3 = 55% 6.3–40 = 35% 40–60 = 50% >60 = 25%
E. coli (EC)	NF EN ISO 9308−3 (March 1999)	56 MPN/100 mL 38 MPN/100 mL 15 MPN/100 mL	Undetermined
Intestinal enterococci (IE)	NF EN ISO 7899−1 (March 1999)	56 MPN/100 mL 38 MPN/100 mL 15 MPN/100 mL	Undetermined
Total suspended solids (TSS)	NF EN 872 (June 2005)	2 mg/L	2–6 = 60% >6 = 20%
Nitrate	NF EN ISO 13395 (October 1996)	0.4 mg N/L	0.4–1.4 = 35% >1.4 = 10%
	NF EN ISO 10304−1 (July 2009)	0.5 mg NO3/L	0.5–1.5 = 30% >1.5 = 10%
Nitrite	NF EN ISO 13395 (October 1996)	0.02 mg N/L	0.02–0.09 = 45% >0.09 = 10%
	NF EN ISO 10304−1 (July 2009)	0.01 mg NO2/L	0.01–0.4 = 40% >0.4 = 10%
Orthophosphates (PO$_4^{3-}$)	NF EN ISO 15681−2 (June 2005)	0.1 mg P/L	0.1–0.4 = 40% >0.4 = 10%
	NF EN ISO 15681−2 (June 2005)	0.02 mg PO4/L	0.02–0.08 = 40% >0.08 = 10%

pH	NF EN ISO 10523 (May 2012)	–	10%
Total phosphorus (Pt)	SIAAP internal method	0.3 mg P/L	0.3–1.1 = 55% >1.1 = 15%
Spores of sulfite reducing microorganisms (SSR)	NF EN 26461–2 (July 1993)	200 CFU/100 mL 34 CFU/100 mL 10 CFU/100 mL	Undetermined
Sulfate	SIAAP internal method	10 mg/L	10–25 = 25% >25 = 10%
	NF EN ISO 10304–1 (July 2009)	1 mg/L	1–1.5 = 15% >1.5 = 10%
Turbidity	NF EN ISO 7027–1 (August 2016)	0.5 FNU	>0.5 = 30%

Table 12 Disinfection pilot operation installation: measurement summary and abatement calculation.

In Situ Disinfection	ALERT Pre-Disinfection $E.\ coli$/100 mL	Lab Pre-Disinfection MPN/100 mL	ALERT Post-Disinfection $E.\ coli$/100 mL	Lab Post-Disinfection MPN/100 mL	ALERT Removal log_{10}	Lab Removal log_{10}
Minimum value	14,574	10,100	4	15		
Maximum value	26,272	21,600	29	61		
Average	21,947	15,917	8	33	3.4	2.7

SECTION 2 – CHAPTER 3

Table 13 Conventional wastewater quality parameters measured in SEV WWTP discharge before and after PFA disinfection.

	SEV WWTP Discharge (Average ± Standard Deviation)	SEV Disinfected Water (Average ± Standard Deviation)
TSS (mg/L)	5.0 ± 3.9	5.4 ± 3.8
DOC (mgC/L)	6.5 ± 0.5	7.7 ± 0.6
COD (mgO$_2$/L)	22 ± 5	24 ± 5
BOD$_5$ (mgO$_2$/L)	1.9 ± 1.5	2.5 ± 1.6
TKN (mgN/L)	1.4 ± 0.6	1.4 ± 0.5
N-NH$_4^+$ (mgN/L)	0.4 ± 0.2	0.3 ± 0.2
N-NO$_3^-$ (mgN/L)	14.6 ± 2.1	15.2 ± 2.1
N-NO$_2^-$ (mgN/L)	0.05 ± 0.04	0.04 ± 0.03
TP (mgP/L)	0.9 ± 0.5	0.9 ± 0.4
P-PO$_4^{3-}$ (mgP/L)	0.6 ± 0.5	0.6 ± 0.3
pH	7.07 ± 0.11	7.26 ± 0.09
Conductivity (µS/cm)	1100 ± 121	1103 ± 122

Table 14 Effect of PFA applied to SEV WWTP discharge at the industrial scale on *E. coli*, intestinal enterococci, SSR and F-specific RNA phages.

Date	PFA Dose ppm	PFA c × t ppm. min	TSS Mg/L	*Escherichia Coli*			Intestinal Enteroccoci		
				Upstream Injection MPN/100 mL	Downstream Injection MPN/100 mL	Log Removal Log	Upstream Injection MPN/100 mL	Downstream Injection MPN/100 mL	Log Removal log
28/08/18	1.0	31.2	3	10,000	76	2.12	1700	<15	≥2.05
11/09/18	1.2	31.6	3	18,600	353	1.72	5030	<15	≥2.53
25/09/18	1.2	36.4	9	51,700	504	2.01	11,500	292	1.60
09/10/18	0.8	22.1	3	15,800	612	1.41	2960	232	1.11
23/10/18	2.2	73.8	3	10,100	46	2.34	4420	<15	≥2.47

Date	PFA Dose ppm	PFA C × t ppm. min	TSS Mg/L	Spore of Sulfite-Reducing Anaerobes			F-Specific RNA Phages		
				Upstream Injection CFU/100 mL	Downstream Injection CFU/100 mL	Log Removal Log	Upstream Injection PFU/100 mL	Downstream Injection PFU/100 mL	Log Removal log
28/08/18	1.0	31.2	3	1170	855	0.14	<30	<30	
11/09/18	1.2	31.6	3	1530	1200	0.11	<30	<30	
25/09/18	1.2	36.4	9	6000	7000	NR	170	<30	≥0.75
09/10/18	0.8	22.1	3	883	2700	NR	<30	<30	
23/10/18	2.2	73.8	3	1650	2000	NR	<30	<30	

NR = Not relevant.

Table 15 Summary of the different variants considered during model construction.

Dataset Used	Explained Variable	Explanatory Variables	Interactions	n
Full dataset	$\ln(EC_{out})$	Q, H, T_{plant}, PFA dose, Ct, $pH_{sampling}$, $T_{sampling}$, $\sigma_{sampling}$, EC_{in}, $[TSS]_{in}$, $[DOC]_{in}$, $[COD]_{in}$, $[BOD]_{in}$, $[TKN]_{in}$, $[NH_4^+]_{in}$, $[NO_2^-]_{in}$, $[NO_3^-]_{in}$, $[Ptot]_{in}$, $[PO_4^{3-}]_{in}$, pH_{lab}, σ_{lab}, turbidity	None Two-way	48
'Online' dataset		Q, H, T_{plant}, PFA dose, Ct, $pH_{sampling}$, $T_{sampling}$, $\sigma_{sampling}$, EC_{in}, $[DOC]_{in}$, $[NH_4^+]_{in}$, pH_{lab}, σ_{lab}, turbidity	None Two-way	

Table 16 Main results from model construction on the full dataset.

ID	Source	Variables Included	p	Maximum VIF Value	R^2	Adj. R^2	MSE	MSE_{val} (LOO)	MSE_{val} (5-fold)
F1	Forward SSE and BIC	Dose, Ct, $T_{sampling}$, $[NH_4^+]$	4	23.1	0.826	0.810	0.704	1.401	1.429
F2	Forward AIC	Dose, Ct, $T_{sampling}$	3	23	0.793	0.779	0.818	1.726	1.740
F3	Backward SSE, BIC and AIC1	Q, H, T_{plant}, Dose, [DOC], [COD], [BOD], $[NH_4^+]$	8	11.6	0.883	0.859	0.511	0.730	0.762
F4	Backward AIC2	Q, H, T_{plant}, $pH_{sampling}$, Dose, [DOC], [COD], [BOD]	8	11.6	0.885	0.862	0.523	0.807	0.857

Table 17 Main results from model construction using the 'online' dataset.

ID	Source	Variables Included	p	Maximum VIF Value	R^2	Adj. R^2	MSE	MSE_{val} (LOO)	MSE_{val} (5-fold)
S1	All forward procedures	Dose, Ct, $T_{sampling}$, $[NH_4^+]$	4	23.1	0.826	0.810	0.704	1.401	1.429
S2	All backward procedures	Q, H, T_{plant}, Dose, [DOC], $[NH_4^+]$	6	10.5	0.847	0.825	0.667	1.023	1.049

Table 18 Main results from model construction on the 'online' dataset with interactions.

ID	Source	Variables Included	P	Maximum VIF Value	R^2	Adj. R^2	MSE	MSE_{val} (LOO)	MSE_{val} (5-fold)
SI1	Forward AIC	Q, H, $T_{sampling}$, [DOC], Dose*$T_{sampling}$, Dose*[DOC]	6	147 (10.4)	0.871	0.852	0.548	0.746	0.763
SI2	Backward SSE	Q, $T_{sampling}$, Dose, [DOC], H*$T_{sampling}$	5	12.2 (10.6)	0.844	0.825	0.647	1.155	1.177

Table 19 Concentrations of AOX, bromide and bromate in SEV WWTP discharge before and after PFA disinfection.

Date	PFA Dose ppm	PFA $C \times t$ ppm. min	TSS Mg/L	A.O.X.			Bromide			Bromate		
				Upstream Injection $\mu gCl/L$	Downstream Injection $\mu gCl/L$	Variation %	Upstream Injection $\mu gBr/L$	Downstream Injection $\mu gBr/L$	Variation %	Upstream Injection $\mu gBrO_3/L$	Downstream Injection $\mu gBrO_3/L$
28/08/18	1.0	31.2	3	86	110	+28	100	90	−10	<5	<5
11/09/18	1.2	31.6	3	39	48	+23	120	<50	−58	<5	<5
25/09/18	1.2	36.4	9	52	57	+10	250	160	−36	<5	<5
09/10/18	0.8	22.1	3	57	73	+28	170	170	0	<5	<5
23/10/18	2.2	73.8	3	68	96	+41	170	170	0	<5	<5

SECTION 3 – CHAPTER 1

Table 20 Biological panels of general toxicity applied during this project.

Panel	Biological Models	Observed Effects	Number of Tests
General toxicity	Bacterial (two *E. coli* strains)	Growth (OD)	2
	Yeast – Cryptogams eukaryotes (*Saccharomyces cerevisia*)	Growth (OD)	2
	Fungal – Ascomycete (*Septoria tritici*)	Growth (OD)	1

Table 21 Results obtained for the various biological models of general toxicity applied in this project.

		Strong Effect	Moderate Effect	Weak, Yet Significant, Effect	No Significant Effect	Weak, Yet Significant, Effect	Moderate Effect	Strong Effect
Bacterial	AG100A	Below −90%	−30 to −90%	−15 to −30%	−15 to +15%	15–30%	30–90%	Above 90%
	NR698	Below −90%	−60 to −90%	−30 to −60%	−30 to +30%	30–60%	60–90%	Above 90%
Yeast	WT	Below −90%	−60 to −90%	−30 to −60%	−30 to +30%	30–60%	60–90%	Above 90%
	AD1−9	Below −90%	−30 to −90%	−15 to −30%	−15 to +15%	15–30%	30–90%	Above 90%
Fungal	*Septoria tritici*	Below −95%	−60 to −95%	−30 to −60%	−30 to +30%	30–60%	60–95%	Above 95%

Table 22 Endocrine disruption panels applied in this project.

Panel	Biological Models	Observed Effects
Endocrine disruption	Amphibian xenopus (*Xenopus laevis*) TH/bZIP Fish larvae medaka ChgH-GFP (*Oryzias latipes*) Fish larvae medaka SPG-GFP (*Oryzias latipes*)	Thyroid activity (effect on development) Estrogenic activity (effect on reproduction) Androgenic activity (effect on reproduction)

SECTION 4 – CHAPTER 2

Table 23 Synthesis of the main laboratory and full-scale studies conducted over the period 2005–2018.

Disinfectant	Years	Plants/Matrices	Investigations
PFA	2005/2006, 2011, 2012, 2013–2018	CL-ER-JS-SD	Effectiveness; QI_1; QI_2; ecotoxicity; genotoxicity; decomposition/residuals
PAA	2002, 2006, 2008, 2013, 2017	JS-SD	Effectiveness; QI_2; decomposition/residuals
HYP	2008, 2011, 2012	CL-ER-JS-SD	Effectiveness; decomposition/residuals

Notes: CL = Caorle; ER = Eraclea; JS = Jesolo; SD = San Donà; QI_1 = qualitative impacts in terms of physicochemical parameters, such as pH, total organic carbon, biochemical oxygen demand and formate; QI_2 = Oxidation power and byproducts; PFA = performic acid; PAA = peracetic acid; HYP = chlorine hypochlorite.

Table 24 Methods for chemical and microbiological investigations conducted at both the full and laboratory scales.

Parameter		Method	Description
Fecal indicators	Fecal coliforms	APAT CNR-IRSA Man 29 2003	MF; mFC agar
	E. coli	(7020B, 7030D, 7040C)	MF; chromogenic *E. coli* agar
	Fecal Enterococci		MF; Slanetz and Bartley/Bile Esculin Azide agar
Quality parameters and impacts	pH	APHA ed. 21st 2005–22nd 2012	Electrometric method
	TOC	(4500-H+B, 5310 C, 5210 D)	Persulfate-Ultraviolet Oxidation Method
	BOD_5		Respirometric method (WTW)
	TSS – Turbidity	APAT CNR-IRSA Man 29 2003	Gravimetric - Nephelometric
	Ammonium	UNI EN ISO 14911:2001	Ion chromatography
	Nitrite	UNI EN ISO 10304–1:2009	Ion chromatography
	COD	UNI EN ISO 15705:2002	Photometric measurement in sealed tube
	Formate	EPA 300.1 1999	Ion chromatography
Ecotoxicity	Vibrio Fischeri	UNI EN ISO 11348–3B 2001–2009	Microtox test system for determining light emission suppression
	Daphnia Magna Straus	UNI EN ISO 6341:2013	Inhibition of the mobility of Daphnia Magna Straus (cladocera, crustacea)

Notes: MF = Membrane filtration; TOC = total organic carbon; BOD5 = biochemical oxygen demand; TSS = total suspended solids; COD = chemical oxygen demand.

Table 25 Full-scale experimental operating conditions – average (minimum–maximum).

Disinfection System	Year	WWTP	Flow Rate (m³/d)	Disinfectant Dose (mg/L)	Retention Time (min)
PFA-HP prototype	2005–2006	CL	7140 (5400–10,000)	1.1 (0.3–1.8)	24 (10–45)
Pilot	2006		10,900 (3410–20,100)	1.6 (0.9–2.4)	36 (16–59)
PFA	2011	ER	3190 (1790–4970)	1.0 (0.6–1.2)	11 (6–19)
	2013–2018	CL-ER-JS-SD	Typical	0.6 (0.4–1.1)	26 (5–83)
PAA	2006	JS	32,570 (23,620–43,680)	1.4 (0.7–2.5)	27 (13–49)
HYP	2008	CL	10,110 (6860–16,350)	2.6 (1.9–4.7)	24 (12–51)
	2011	JS	24,120 (15,920–34,520)	2.7 (1.9–3.2)	20 (15–30)

Notes: CL = Caorle; ER = Eraclea; JS = Jesolo; SD = San Donà.

Table 26 Median and range of variation of the main chemical-physical parameters of effluents entering the disinfection treatment during the full-scale studies performed from 2005 to 2011.

	pH (unit)	TSS (mg/L)	COD (mg/L)	N-NH$_4$ (mg/L	F. coliforms (Log)	*E. coli* (Log)	Enterococci (Log)
PFA-HP	7.6 (6.9–8.2)	11.9 (9.0–34)	25 (10–35)	0.57 (0.02–3.7)	4.0 (2.3–5.3)	3.6 (2.0–5.3)	3.4 (2.7–4.5)
PFA	7.7 (7.1–8.3)	9.9 (9.0–17)	26 (14–45)	1.5 (0.02–9)	4.9 (4.4–6.0)	4.6 (3.0–5.6)	3.4 (2.7–4.3)
PAA	7.4 (7.1–8.0)	10 (9.0–11)	34 (14–47)	0.14 (0.14–3.7)	5.2 (4.7–5.3)	4.4 (3.3–5.2)	3.4 (2.7–4.3)
HYP	7.5 (7.3–7.8)	9.0 (9.0–17)	29 (21–48)	1.94 (0.14–10)	5.3 (5.0–6.0)	4.6 (3.6–6.2)	3.5 (2.6–4.6)

Notes: HYP = hypochlorite; PAA = peracetic acid; PFA-HP = Hyproform performic acid solution; PFA = Desinfix performic acid solution; TSS = total suspended solids; COD = chemical oxygen demand; F. coliforms = fecal coliforms.

Table 27 Average and range of variation of bacterial inactivation achieved by PFA, PAA and HYP for each interval of contact time (Kruskal–Wallis test) during the full-scale studies performed from 2005 to 2011.

Time (min)	Disinfectant Type	Doses (mg/L)	Ret. Time (min)	E. coli (log R)	Enterococci (log R)	F. coliform (log R)
<=10	PFA	1.04 ± 0.07	8.32 ± 1.12	2.92 (2.23–3.97)	1.75 (0.65–2.78)	–
>10–20	PFA-HP	1.26 ± 0.54	14.38 ± 2.24	1.05 (0.06–3.28)[c]	0.76 (0.46–2.31)[b]	1.02 (0.30–3.75)[b]
	PFA	0.88 ± 0.22	13.26 ± 2.67	3.3 (2.04–4.19)[a]	1.78 (0.67–3.15)[a]	
	PAA	1.60 ± 0.38	16.16 ± 1.96	1.92 (1.16–2.61)[b]	0.35 (0.04–0.92)[c]	1.57 (0.84–2.69)[b]
	HYP	2.83 ± 0.70	16.34 ± 2.22	2.71 (0.98–4.05)[b]	1.19 (0.36–3.09)[ab]	3.45 (2.11–4.28)[a]
>20–30	PFA-HP	1.16 ± 0.57	22.49 ± 3.25	1.87 (0.32–3.65)	1.42 (0.42–3.26)[a]	1.67 (0.38–3.32)
	PAA	1.94 ± 0.67	25.12 ± 3.32	2.86 (0.16–3.42)	0.28 (0.14–0.71)[b]	–
	HYP	2.45 ± 0.47	25.39 ± 2.89	2.75 (1.21–5.15)	1.48 (0.66–2.23)[a]	–
>30	PFA-HP	1.35 ± 0.60	45.07 ± 10.24	2.52 (0.44–4.29)	1.92 (0.55–3.67)[a]	2.57 (0.62–4.76)
	PAA	1.07 ± 0.20	38.94 ± 6.77	2.45 (1.21–3.37)	0.46 (0.1–1.63)[b]	–
	HYP	2.38 ± 2.46	36.54 ± 7.16	2.51 (1.78–3.13)	1.75 (0.63–3.12)[a]	–

Notes: Ret. Time = retention time; log R = log reduction (CFU/100 mL); PFA-HP=Hyproform performic acid solution; PFA = Desinfix performic acid solution; F. coliforms = fecal coliforms; different superscript letters (down columns) mean significant different variations at $p = 0.05$, with values[a] > values[b] > values[c].

Table 28 Order of effectiveness of the disinfectant against fecal indicators from 2005 to 2011 by the ANCOVA test.

F. coliforms log R (CFU/100 mL)	E. coli log R (CFU/100 mL)	Enterococci log R (CFU/100 mL)
Comparable ($p > 0.05$)	PFA > PAA~PFA-HP > HYP	PFA > PFA-HP > HYP~PAA

Notes: F. coliforms = fecal coliforms; log R = log reduction (CFU/100 mL); PFA-HP = Hyproform performic acid solution; PFA = Desinfix performic acid solution.

Table 29 Disinfection kinetics parameters estimated for the hom and S models, minimum and maximum values (dose range: PFA = 0.5–1.3 mg/L – PAA = 1–3 mg/L).

		E. coli		Enterococci	
		PFA	PAA	PFA	PAA
HOM	n	0.000–0.416	0.000–0.010	0.000–0.401	0.000–1.181
	m	0.000–0.491	0.001–1.141	0.353–0.774	0.225–1.809
	K	2.138–8.928	0.152–8.202	0.305–1.743	0.002–1.212
	R^2	0.969–1.000	0.917–1.000	0.943–0.988	0.922–0.998
S Model	n	0.000–0.139	0.000–0.160	0.001–2.877	0.537–0.598
	m	0.020–8.794	3.059–24.561	0.916–3.524	0.633–4.671
	h	0.006–9.587	5.538–16.251	10.200–30.522	5.500–53.00
	K	6.217–17.571	4.303–8.294	6.011–37.479	2.296–4.457
	R^2	0.984–1.000	0.969–1.000	0.930–0.999	0.943–1.000

Notes: PFA = Desinfix performic acid solution.

Table 30 Percentage of results being in compliance with various levels of microbiological target at disinfection outlet and correspondent disinfection operating conditions (average and range of variation).

WWTP	Disinfectant	Dose (mg/L)	Retention Time (min)			
CL	PFA-HP	1.3 (0.3–2.4)	28 (10–59)			
ER	PFA	1.0 (0.6–1.2)	11 (6–19)			
JS	PAA	1.6 (0.9–2.1)	16 (13–20)			
CL – JS	HYP	2.8 (2.5–3.0)	17 (15–20)			
	CFU/100 mL	<12,000	<2400	<1000	<100	
Fecal	PFA-HP	100%	88%	83%	55%	
coliforms	PFA	–	–	–	–	
	PAA	92%	42%	25%	0%	
	HYP	100%	100%	100%	100%	
	CFU/100 mL	<5000	<1000	<100	<10	
E. coli	PFA-HP	100%	90%	67%	35%	
	PFA	100%	100%	76%	31%	
	PAA	92%	83%	17%	0%	
	HYP	100%	93%	59%	14%	
	CFU/100 mL	<2000	<400	<100	<10	
Enterococci	PFA-HP	88%	78%	48%	15%	
	PFA	100%	90%	69%	10%	
	PAA	67%	17%	0%	0%	
	HYP	95%	90%	37%	10%	

Notes: CL = Caorle; ER = Eraclea; JS = Jesolo; Ret. Time = retention time; PFA-HP = Hyproform performic acid solution; PFA = Desinfix performic acid solution.

Table 31 Bacterial concentrations (average $\pm$ standard deviation) at the disinfection outlet under the disinfectant operating conditions of Table 30.

	E. coli CFU/100 mL	Enterococci CFU/100 mL	F. coliform CFU/100 mL
PFA	62 ± 81[b]	169 ± 153[b]	454 ± 429[b]
PAA	1143 ± 201[a]	1757 ± 375[a]	5276 ± 964[a]
HYP	643 ± 155[a]	1311 ± 310[a]	2784 ± 1609[a,b]

Notes: F. coliforms = fecal coliforms; different superscript letters (down columns) mean significant different variations at $p = 0.05$ with values[a] >values[b] > values[c].

Table 32 Percent of results, recorded between 2013 and 2018, meeting the various *E. coli* targets; percentages are expressed for each contact time interval and the corresponding average applied PFA.

Retention Time (min)	Samples No.	% Samples Respecting the *E.coli* Limits (CFU/100 mL) (%)				PFA Average Dose mg/L
		<1000	<100	<10	>1000	
≤10	56	96%	48%	0%	4%	0.7
10–20	127	98%	72%	9%	2%	0.6
>20–30	65	100	31	52		0.6
>30–40	33	100	27	48		0.6
>40	77	100	55	6		0.6
Total	358	99	53	18	1%	0.6

Table 33 Qualitative parameter variation at the disinfection inlet from 2013 to 2018; data are reported as average, standard error and (maximum value).

	TSS (mg/L)	COD (mg/L)	N-NH$_4$ (mg/L)	N-NO$_2$ (mg/L)
Caorle	10 ± 6.4 (36)	19 ± 6.8 (44)	0.58 ± 1.12 (7.0)	0.13 ± 0.13 (0.57)
Eraclea	9.3 ± 4.4 (35)	21 ± 6.4 (54)	1.35 ± 2.18 (17.9)	0.24 ± 0.22 (1.02)
Jesolo	10.0 ± 6.0 (33)	19 ± 7.2 (57)	1.85 ± 1.96 (7.8)	0.12 ± 0.10 (0.67)
San Donà	8.7 ± 3.6 (27)	19 ± 6.4 (45)	0.31 ± 0.57 (4.43)	0.06 ± 0.05 (0.28)

Notes: TSS = total suspended solids; COD = chemical oxygen demand.

Table 34 Comparisons between TOC at the dosage point (T_0) and disinfection outlet (t-student test) for each time interval – average ± standard error (range of variation).

Time Interval (min)	PFA Dose (mg/L)	TOC T_0 (mg/L)	TOC OUT (mg/L)	Probability
<10	1.01 ± 0.18	6.79 ± 0.63 (5.34–8.0)	6.94 ± 0.64 (5.40–7.8)	0.116
10–20	0.87 ± 0.83	6.12 ± 1.05 (4.47–8.95)	6.24 ± 0.97 (4.75–8.90)	0.032
20–30	0.89 ± 0.46	7.19 ± 1.30 (5.20–9.38)	7.17 ± 1.31 (5.60–9.53)	0.084
>30	1.16 ± 0.60	7.37 ± 1.13 (5.52–9.41)	7.57 ± 1.19 (5.28–10.26)	0.001

Notes: TOC = total organic carbon; OUT = disinfection outlet.

Table 35 Comparisons between formate at the dosage point (T_0) and disinfection outlet (Wilcoxon rank test) for each time interval; median and range of variation.

Time Interval (min)	Formate T_0 (mg/L)	Formate OUT (mg/L)	Probability
<10	2.81 (1.62–3.6)	3.10 (1.97–3.6)	0.061
10–20	2.04 (1.35–10.89)	2.0 (0.82–11.08)	0.168
20–30	3.51 (1.15–9.77)	3.08 (1.15–9.92)	0.079
>30	4.04 (1.45–10.31)	3.97 (1.40–10.50)	0.524

Notes: OUT = disinfection outlet.

Table 36 Carbon components associated with PAA and PFA-HP, measured at the laboratory in batch tests by dosing the disinfectant and/or the associated carbon ions in the disinfection mixture.

Disinfectant Type	Doses (mg/L)	Increase for Each mg/L of Active Substance		
		TOC (mg/L)	COD (mg/L O2)	BOD$_5$ (mg/L O2)
PFA-HP	2–50 Formate ion	1.3 ± 0.13	1.8 ± 0.13	0.5 ± 0.21
PAA	1–10 PAA	0.9 ± 0.03	–	1.1 ± 0.18
	2–50 Aceate ion	0.9 ± 0.12	2.0 ± 0.17	1.6 ± 0.29

Notes: PFA-HP = Hyproform performic acid solution.

Table 37 Percentage of cases in which the two peracids respected the microbiological target set at 20% of the guideline or limit values; doses and retention times refer to the conditions under which the target was respected.

Disinfectant	Fecal Indicator	Cases Respecting Target (%)	Doses (mg/L)	Cases at Retention Time <20 min	Cases at Retention Time >=20 min
PFA	E.coli	99%	0.9	60%	40%
PAA		92%	2.4	30%	62%
PFA	Enterococci	88%	0.8	53%	35%
PAA		48%	2.8	7%	41%

IWA Publishing's authorised EU representative for General Product
Safety Regulations is Diane D'Arras, 15 rue Duret, 75116 Paris,
France, e-mail: safety@iwap.co.uk.

Printed and bound by CPI Group (UK) Ltd, Croydon, CR0 4YY
16/06/2026
02140355-0001